The Eureka Factor

Creative Insights and the Brain

John Kounios & Mark Beeman

1 3 5 7 9 10 8 6 4 2

Windmill Books
20 Vauxhall Bridge Road
London SW1V 2SA

Windmill Books is part of the Penguin Random House group of companies whose addresses can be found at global.penguinrandomhouse.com.

First published in the United States by Random House, an imprint and division of Penguin Random House LLC, New York in 2015

First published in Great Britain by William Heinemann in 2015
First published in paperback by Windmill Books in 2016

www.windmill-books.co.uk

A CIP catalogue record for this book is available from the British Library.

ISBN 9780099537373

Illustrations on pages 18, 24, 59-61, 81, 86 and 133 are by Sharon O'Brien, and illustrations on pages 35 and 176 are by Casey Hampton.

Book design by Casey Hampton

Printed and bound by CPI Group (UK) Ltd, Croydon, CR0 4YY

The Eureka Factor

Dr Kounios' and Dr Beeman's insight research has been reported by *The Times*, *The New Yorker*, the *New York Times*, the *Wall Street Journal*, *BBC's Horizon*, National Public Radio, *Scientific American Mind*, *O*, *The Oprah Magazine* and other international print and electronic media. Their work is displayed in an exhibit in the Museum of Science and Industry in Chicago.

PREFACE

"Eureka!" No one knows for sure whether Archimedes really shouted this word, jumped from his bathtub, and ran through the streets of ancient Syracuse proclaiming his latest discovery. But the story has persisted for two millennia because it resonates with people—you have probably had such "aha moments" or sudden realizations yourself. These "insights," as psychologists call them, are powerful experiences that expand our understanding of the world and ourselves. They can confer both enlightenment and practical advantage.

Stories of insight resonate with us, as well, which is why we've been studying these moments for almost twenty years. It's why we wrote this book. Our goal is to explain what insights are, how they arise, and what the scientific research says about how to have more of them. But first we would like to tell you a bit of the history of our work and, more generally, of research on insight.

During the decades following World War I, German psychologists documented that when faced with a confusing and seemingly intractable problem, a person may suddenly realize that he or she had been thinking about it in the wrong way and that the solution is actually quite straightforward. Solving a problem is all about how you "see" it.

After identifying insight, psychologists focused on characterizing it. In particular, they sought to show that it is unique and different from deliberate, conscious thought—what they call "analysis." For example, in the 1980s, psychologist Janet Metcalfe showed that people can consciously monitor their deliberate analytic thought; however, the mental processes leading up to an insight are largely unconscious, making it difficult to monitor them and predict when a solution will burst into awareness as an aha moment. Another advance occurred in the early 1990s when another psychologist, Jonathan Schooler, demonstrated that insightful thought is fragile and easily overshadowed—thinking out loud makes it less likely that you will solve a problem with a flash of insight, but talking your way through a problem won't impair your ability to solve it analytically.

Despite Schooler's discovery, by the 1990s new research findings about insight were rare. The field had become almost dormant. Though insight remained a core topic of experimental psychology covered in nearly every introductory psychology textbook, no one had been able to pin down its mechanics. The most important questions remained: How do insights occur? Can we spur more of them?

There was an obstacle to progress, one that has to do with insight's very essence: It *feels* different. The potency of aha moments is why people notice and remember them. Nevertheless, some skeptics maintained that this feeling is misleading and that insights differ from deliberate thought *only* in how people feel when they reach a solution. Otherwise, they are nothing special. Eurekas as true creative breakthroughs, they argued, are fairy tales.

When we met while working at the University of Pennsylvania in late 2000, we discussed whether the skeptics could be right. What if aha moments feel different but aren't otherwise unique? Perhaps they are just ordinary thought that occasionally yields extraordinary results. If only there were some objective marker to validate the subjective experience of insight—something that would help us to isolate aha moments and analyze them to figure out whether they are distinct.

We realized that this kind of objective marker of insight does potentially exist—in the activity of the brain. That set us on our path.

Until then, Mark's research had focused on a different topic: how language comprehension relies on the brain's right hemisphere—the side of the brain noted more for spatial processing than for language. Based on others' research and his own studies of subtle language deficits in patients with damaged right hemispheres, he proposed a theory of how the hemispheres process information differently from each other. A turning point in Mark's career occurred in 1994 when he heard Jonathan Schooler lecture about insight. This convinced him that the same characteristic of the right hemisphere that enables people to flexibly comprehend language—namely, the ability to draw together distantly related information—also contributes to aha moments. During the 1990s, Mark teamed up with Edward Bowden, an insight researcher he knew from their graduate school days, to collaborate on behavioral studies that provided support for the special role of the right hemisphere in insight. Meanwhile, Mark began investigating language with fMRI—functional magnetic resonance imaging—to map out the brain areas that enable people to comprehend stories. Soon, he began to think about using fMRI to study insight.

In the 1990s, John's main research interest was the neural basis of "semantic memory"—how people acquire, use, and sometimes lose their knowledge. He recorded the brain's electrical activity with EEG—electroencephalography—to trace out, moment by moment, how one brings a concept to mind. Looking at how insights spring to mind was the next logical step. He and Roderick Smith, his doctoral student, published a behavioral study showing that insights arise abruptly and in their entirety, validating the conscious experience of suddenness. This started him thinking about using EEG to study insight.

The field of brain imaging started to take off in the early 1990s and developed rapidly throughout that decade. The availability of these techniques meant that we weren't limited to observing the outward behavior of people. We could peer inside their working brains. That changed everything.

Early neuroimagers mostly investigated abilities that had already been extensively explored by psychological scientists, such as perception, attention, movement, and memory. They shied away from the more difficult task of investigating mental abilities that were more complex and less well understood, such as reasoning, decision making, and problem solving—never mind insight.

We believed that we were ready to use these tools to study insight, but we had a scientific decision to make: Which experiment should we run? Research funds and time were scarce. Each of us had just enough funding to support one experiment. But which experiment should we do? We circled around this question for a while but kept coming back to the one issue that proved to be the key: What happens in the brain at the instant when a person solves a problem with a flash of insight? We designed an experiment that would illuminate the aha moment itself.

By 2002 we had hammered out the details for our first study and were ready to start testing. However, we were anxious because we were taking a big chance. Ideally, researchers like to run small preliminary "pilot" experiments to work out the kinks so they can refine their procedures before running a full study. We didn't have the resources or the time to do that, so we had to get it right on the first shot.

After collecting the data, we spent the next few months independently analyzing the EEG and fMRI results. Then we traded our brain images and were astounded—the EEG and fMRI images, when superimposed, formed a nearly perfect match! The main result: A key area of the right hemisphere lights up at the aha moment. This and other findings finally provided concrete evidence for the reality and distinctiveness of insight. We began preparing an article describing our results. By the time we had moved on to new faculty positions—Mark to Northwestern University and John to Drexel University—we submitted it for publication. We were delighted when it was accepted by the prestigious journal *PLoS Biology*.

That paper attracted a great deal of attention from fellow psychologists and neuroscientists. Researchers had always maintained an interest in insight, even when there had been little new evidence to fuel this fascination. But we did not anticipate the extent of the positive reaction from the news media and the public. For example, *The Times* of London enthusiastically proclaimed the discovery of the brain's "E-spot" ("E" for "eureka"), a simplification necessary for boiling down our findings. This kind of coverage spurred people from all walks of life to send us letters and emails describing their own aha moments and personal intuitions about creativity. Some of these stories have found their way into this book. Others have inspired new experiments.

That first neuroimaging study suggested further research, which we have continued to this day as the main focus of our work. As this research progressed, it became clear that the emerging story of insight could not be told in a newspaper article. It would take a book.

We set out to write one that was both lively and readable. Just as important, we wanted to ensure its scientific accuracy; documentation of this, along with interesting information from the cutting-room floor, can be found in the notes. We also wanted to both evoke the wonder of discovery and inspire people to use the research to be more creative in their personal and professional lives. To help achieve these goals, we have included many anecdotes that illustrate aha moments and the circumstances that led to them. As scientists, we don't consider anecdotes to be definitive evidence for or against scientific theories, because any single anecdote could be an exceptional case or misreported. But they do help to illustrate key ideas. They have also inspired us and, we think, will inspire you.

Writing this book has been a tremendous experience. But the real gratification for us comes from sharing this information. We hope that it will empower you to use creative insight to realize and surpass your personal and professional aspirations.

CONTENTS

The Eureka Factor

1

NEW LIGHT, NEW SIGHT

But who can count or weigh such lightning flashes of the mind? Who can trace out the secret threads by which our conceptions are united?

—Hermann von Helmholtz, scientist

Helen Keller didn't know what a word was. When she was nineteen months old, a brief illness left her permanently deaf and blind, preventing her from learning to speak. Eventually, she developed a few signs for basic communication, but they were just gestures. She was imprisoned within a world of palpable objects. The realm of words and ideas was beyond her grasp.

In 1887, when Helen was six years old, her parents hired a young teacher named Anne Sullivan to tutor her at home. Anne, who became Helen's lifelong friend and companion, attempted to teach Helen words by tracing them on her young student's palms. Helen learned several tracings this way, but she wasn't able to comprehend that they were words. "I did not know that I was spelling a word or even that words existed; I was simply making my fingers go in monkey-like imitation," she later explained.

One day, Helen and Anne had a tussle over the words "mug" and "water." Helen couldn't connect the tracings with their respective objects. At a later lesson, she became upset and smashed her doll. Anne tried a different approach. She took Helen to the well house and directed her to hold her mug under the spout while Anne pumped water. As the water poured over Helen's mug and hand, Anne traced the letters "w-a-t-e-r" on Helen's other hand. That's when it happened. According to Anne, "The coming so close upon the sensation of the cold water rushing over her hand seemed to startle her. She dropped the mug and stood as one transfixed. A new light came into her face." As Helen later explained, "I stood still, my whole attention fixed upon the motions of her fingers. Suddenly I felt a misty consciousness as of something forgotten—a thrill of returning thought; and somehow the mystery of language was revealed to me. I knew then that "w-a-t-e-r" meant the wonderful cool something that was flowing over my hand. That living joy awakened my soul, gave it light, hope, joy, set it free!"

In that amazing instant, Helen realized that the scribbles on her hand represented objects in the world and that she could use these symbols to think and to communicate with others. "I left the well-house eager to learn. Everything had a name, and each name gave birth to a new thought. As we returned to the house every object which I touched seemed to quiver with life. That was because I saw everything with the strange, new sight that had come to me."

Thus, a blind girl came to "see."

Helen ultimately learned to read Braille and to write. She learned to speak, even though she couldn't hear, and to read lips with her hands. She graduated from college and went on to write many books of social and spiritual commentary. Mark Twain, Alexander Graham Bell, Charlie Chaplin, and other luminaries of the day befriended her. President Lyndon Johnson awarded her the Presidential Medal of Freedom. She continues to inspire generations to hope and to achieve.

All of this was empowered by a moment of insight.

"Things just clicked." "Everything just snapped into place." "It was a spark of inspiration . . . a bolt of lightning . . . a flash of insight." "Like a lightbulb turning on." "I had an epiphany." "Suddenly, I saw things in a new light."

These expressions all refer to what is commonly called a eureka or an aha moment and what psychologists call "insight" and consider to be a form of creativity. It's the sudden experience of comprehending something that you didn't understand before, thinking about a familiar thing in a novel way, or combining familiar things to form something new. Insights are quantum leaps of thought, creative breakthroughs that power our lives and our history. Insight conveyed a theory of gravity to Sir Isaac Newton, the melody of a Beatles ballad to Sir Paul McCartney, and an understanding of the cause of human suffering to the Buddha. Nearly everyone has had aha moments of sudden clarity. They can and do change our lives.

Much has been written purporting to explain how insight works and how you can make it work better. Almost all of it is based on opinions and informal observations rather than on scientifically established facts. However entertaining or inspiring those popular writings may be, science has now gone much further than anecdotal musings can take us. It's not that opinions and observations are bad. They can be a helpful starting point for inquiry. But there is a more complete approach—a scientific approach. Science finishes the job by putting opinions and observations to the test wherever possible.

Individual fields of science have had periods of extraordinary development, often spurred by new technologies. Astronomy was energized by the invention of the telescope, as biology was by the microscope. The last quarter century has seen the emergence of a new field—cognitive neuroscience—fueled by techniques for measuring the activity of a brain while it works. Techniques such as functional magnetic resonance imaging (fMRI) and high-density electroencephalography (EEG) have enabled us to explore the brain in ways that elucidate how we perceive, remember, think, feel—and have insights. We and our colleagues have used these brain-imaging techniques for more than a decade to uncover what happens in the brain when a person has an aha moment. Combined with the behavioral research methods of cognitive psychology, brain-imaging studies have revealed new and unexpected aspects of insight that would not have been apparent from measuring a person's behavior alone.

We had two aims in writing this book. The first was to explain, based on the latest research from cognitive neuroscience and psychology, exactly what insight is and how it works in the brain. The second was to show you how to use this information to enhance your own creativity and problem solving. These goals are closely intertwined. Media reports, some inaccurate, have trumpeted new research about various factors thought to enhance creativity: Relax, take a vacation, look at the color blue, and so forth. Indeed, there are strategies that will enhance creativity. But these strategies work only when applied correctly—at the right time and in the right context—and the only way to apply them correctly is to grasp how they influence the way you think. Haphazard changes made without comprehension could cause the opposite of what was intended. Our goal is to provide a scientifically based understanding that will enable you to realize your creative potential—at home, at work, at large. In particular, much can be learned by considering how people who tend to experience many insights—we call them "Insightfuls"—think, and how they differ from "Analysts," who tend to rely more on deliberate, methodical thought.

A MATTER OF INTERPRETATION

Before we proceed, we should explain more precisely what we mean by "insight." The word is slippery because it's used to describe a variety of related things. Most often, people use the term to refer to any type of deep understanding, especially of oneself. However, for psychological scientists, insight is more specific and intricate.

Insights have two key features. The first one is that they pop into your awareness, seemingly out of nowhere. They don't feel like a product of your ongoing thoughts. In fact, you can't control them in the way you can control your deliberate, conscious thought. Insights are like cats. They can be coaxed but don't usually come when called.

The other key feature of insights is that they yield, often literally, a different way of looking at things.

Consider the cube on the left side of figure 1.1.

This is a Necker cube. The interesting thing about it is that its appearance is ambiguous. As you can see in the right side of the figure, either the lower square or the higher square of this transparent cube could be viewed as closer to you. With a shift of attention, you can see it in either of these two ways. But you can't see it in both ways simultaneously because the two interpretations are incompatible: a single face of the cube can't be both closer to you and farther from you at the same time. And when you shift your attention from one of these squares to the other, the change in your interpretation is abrupt. This kind of perspective shift is a prototype for insight.

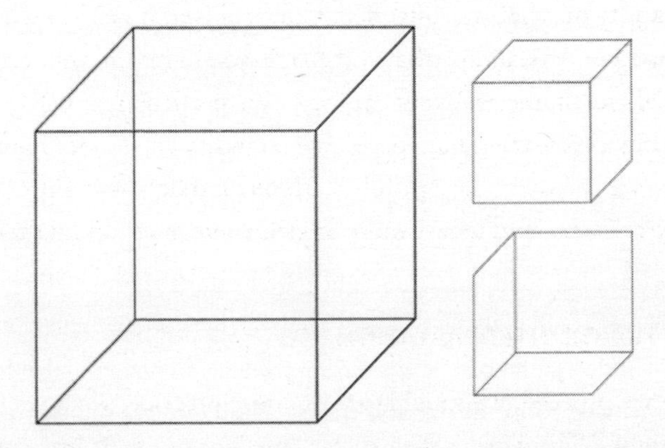

FIGURE 1.1: The Necker cube. *Wikicommons (commons.wikimedia.org/wiki/File:Necker%27s_cube.svg)*

The Gestalt psychologists of the early twentieth century liked to point out that we can interpret almost any type of object, situation, or event in more than one way. That's why people often use expressions such as "seeing things in a new light" or "seeing things from a different angle" to describe insights. If you look at a brick, you'll probably think of it as a part of a building or a wall. But you could also think of it in other ways: as a paving stone, a doorstop, a paperweight, or a walnut cracker. In fact, cognitive psychologists sometimes use the "brick test" as a way to measure creativity: The more frequently you can shift your perspective, the more uses you will be able to think of for a common object such as a brick, and thus the more creative you are considered to be.

According to the Gestalt psychologists, when you get stuck while trying to solve a problem it's often because you are thinking about the problem in the wrong way. Just as a simple visual scene such as a Necker cube can be radically reinterpreted in an instant, so can a complex problem be "restructured," yielding an aha moment about the solution. An object that was previously used for one purpose can now be thought of as a tool to perform some other kind of task; a threat can now be regarded as an opportunity; a relationship with another person can be redefined from competitor to collaborator.

Before Orville and Wilbur Wright's invention of the airplane, the established conception of how powered flight would work was that propellers would produce horizontal thrust by cutting through the air like blades while wings with curved surfaces would provide the airplane with the necessary upward lift. The Wrights had the mental flexibility to shed the old way of thinking about propellers as blades and reimagine them as wings gliding through the air. When they redesigned their propeller blades to give them a curved, wing-like shape—a design feature still used in modern airplanes—the propellers produced more horizontal thrust. This reinterpretation of propellers as rotating wings helped to make powered flight possible.

INSIGHT IS CREATIVE

People often use the terms "insight" and "creativity" interchangeably. Cognitive psychologists, with their penchant for precision, generally consider insight to be a special form of creativity. We'll go a step further and propose that insight is a part of creativity's core.

But then what is creativity? Psychologists often explain it as the ability to generate ideas that are both novel and useful. Though researchers commonly use this definition, we believe that it's inadequate. Creative things do tend to be novel and useful, but what's novel to one person may be reinventing the wheel to someone else. Usefulness is also in the eye of the beholder: An iPhone may be useful to a Manhattan lawyer but would likely be useless to a member of an Amazonian tribe. Moreover, even a truly useless creation can be creative. They're called "brilliant failures."

In the face of a lack of consensus about how to define creativity, one suggestion is to simply not bother defining it—at least not yet. The idea is that everyone intuitively recognizes creativity when he or she sees it and that ongoing research will eventually give birth to a more effective definition. We would argue that this time has come.

We define creativity as the ability to reinterpret something by breaking it down into its elements and recombining these elements in a surprising way to achieve some goal. This understanding covers virtually all of the phenomena that we typically think of as creative. In the hands of a composer, the notes of a musical scale can be rearranged to form a melody. A successful entrepreneur can take well-known components, products, or services and recombine them to produce something that no one else sells and everyone wants to buy. Even creative products that seem radically novel can be seen as a reorganization of familiar elements of perception and thought. The most creative poems, symphonies, paintings, inventions, business plans, or personal realizations are composed of a common reservoir of words, musical notes, colors, parts, processes, steps, or emotions. The basic elements can be familiar. What makes the product creative is how these elements are recombined—the less obvious the recombination, the more creative it is. After all, if it were obvious, then everyone would be doing it.

When this kind of creative recombination takes place in an instant, it's an insight. But recombination can also result from the more gradual, conscious process that cognitive psychologists call "analytic" thought. This involves methodically and deliberately considering many possibilities until you find the solution. For example, when you're playing a game of Scrabble, you must construct words from sets of letters. When you look at the set of letters "A-E-H-I-P-N-Y-P" and suddenly realize that they can form the word "EPIPHANY," then that would be an insight. When you systematically try out different combinations of the letters until you find the word, that's analysis.

Analytic thinking is well suited to familiar situations. When you're trying to form a word in Scrabble or solve an anagram, you know exactly what's available to you, namely, the letters; and you know exactly what you are allowed to do, that is, rearrange them. These things are givens, and if you do enough rearranging, you'll eventually find combinations that make words. Analytic thought is an effective way to deal with such clear-cut problems, but it's less helpful for problems that are too complicated for you to calculate all the permutations or for which it's not entirely clear what you have to work with. For example, if your goal is to be a better parent, find a more rewarding career, or come up with a new idea for a start-up company, then analysis alone may not get you very far. These problems are too fuzzy and complicated for you to methodically evaluate all the possibilities. Moreover, it probably isn't clear what all the tools are that could help you achieve such goals. When you tackle these kinds of problems, insight shines.

EVERYONE, EVERYWHERE

Creative insight is not an exotic type of thought reserved for the few. In fact, it's one of the few abilities that define our species. Animals can to varying degrees do most of the things that humans do—they can see, move, pay attention, and remember. However, except for a few limited and arguable counterexamples, only humans—*most* humans—have insights. It's a basic human ability.

Whenever you suddenly realize how to pay for that new car, why your sibling has distanced herself from you, or how to position yourself for a promotion, you've had an insight. Whenever you are confused by someone's rambling explanation and then, abruptly, understand what he is talking about, you've had an insight. Whenever you are awakened from sleep by the solution to a long-standing problem, you've had an insight. Just about all of us have had such eureka experiences.

Moreover, the products of insight are pervasive. You are surrounded not only by its technological and scientific achievements, such as frozen food and the television, but also by its many imprints on business, the arts, and most other areas of human endeavor. It has even become a staple of popular culture in magazines, self-help books, and on television, which promote the notion that aha moments are a key to one's growth as a person.

The idea that insights are stepping-stones for personal development is not new. Many of the world's great religions have long taught that insights can offer the potential for profound transformation by offering a window onto transcendental and spiritual realms. For example, Exodus 3 describes a familiar biblical epiphany as Moses is suddenly inspired by God to return to Egypt to deliver the Hebrews from Pharaoh's yoke into the Promised Land. Acts 9 recounts the story of Saul of Tarsus (later the Apostle Paul) who was blinded by a light from heaven while on the road to Damascus. His companions led him to Damascus, where "there fell from his eyes as it had been scales: and he received sight forthwith, and arose, and was baptized" (Acts 9:18, King James Version).

The notion of sudden insight plays a special role in Zen Buddhism. The ultimate goal of Zen is "satori," which Japanese scholar D. T. Suzuki explained as "acquiring a new viewpoint for looking into the essence of things." A practitioner may pursue Zen assiduously for many years before achieving satori, but if satori is achieved, it will likely be triggered by some seemingly irrelevant event. According to Suzuki, "An inarticulate sound, an unintelligent remark, a blooming flower, or a trivial incident such as stumbling, is the condition or occasion that will open his mind to satori." He states that this "new viewpoint" will be acquired in an instant: "Satori is the sudden flashing into consciousness of a new truth hitherto undreamed of. It is a sort of mental catastrophe taking place all at once, after much piling up of matters intellectual and demonstrative. The piling has reached a limit of stability and the whole edifice has come tumbling to the ground, when, behold, a new heaven is open to full survey."

A recent survey on spirituality in the United States documents that many people have reported having experienced personally significant religious epiphanies. One question asked people whether they had ever had a "moment of sudden religious insight or awakening"—a question that had been asked in previous surveys. In 1962, when this question was first asked, 22 percent of people surveyed said that they had had such an experience. This figure increased with each survey, reaching 49 percent in 2009. For the first time since the survey started asking this question, more people said that they had had such an epiphany (49 percent) than said that they hadn't (48 percent). Though the survey did not provide an explanation for this increase in reported epiphanies, it is clear that many people believe that they have had such revelations and consider them to be personally meaningful.

Insights are widespread, and people believe them to be important.

But do they really confer any tangible benefits? Let's consider a few areas in which insightful thinking is an asset, and its absence a dilemma.

METAMORPHOSIS

People change over time, often for the better. Maturity, wisdom, patience, and many other strengths can result from the gradual accumulation of life experiences. But do these qualities have to develop slowly? Researcher Timothy Carey and colleagues recently examined the idea that insights can be shortcuts to positive personal change. They conducted structured interviews with people who had just finished psychotherapy. Reports of aha moments abounded. One interviewee said that he could "visualize the point" at which he changed; another said, "I could actually hear it." Many of them could identify the moment at which they had their realizations, such as in a swimming pool with a spouse or in a particular meeting with a therapist. Some used familiar metaphors to describe their ahas, such as a light being turned on, a button being pressed, a click, or a " 'ping' and then it was like I could see things clearly." Personal growth doesn't have to be a glacial process. As physician-author Oliver Wendell Holmes, Sr., wrote, "A moment's insight is sometimes worth a life's experience."

THE INFORMATION VORTEX

According to a recent worldwide survey of more than 1,500 corporate CEOs, the major problem confronting business executives is coping with the increasing complexity of the world. They noted that rapid changes render business models and strategies obsolete almost as soon as they are implemented.

This confusion isn't unique to business—we all face it in every walk of life. As civilization propagates, so, too, do its problems. To the familiar litany of famine, disease, injustice, and war, we can now add newer problems: environmental degradation, increased competition for dwindling natural resources, family fragmentation, cybercrime, and so forth. To make matters worse, these problems have become entwined. Solve the energy shortage by finding and pumping more oil or by building more nuclear power plants? Then how would we deal with increased pollution and toxic waste? Suppress terrorism by tightening security? Then how would we preserve freedom and personal privacy? And so forth. How can we cope with the increasing complexity and interconnectedness of the problems we face?

The CEOs identified creative insight as the way ahead. According to one CEO, in the face of confusion, "It's insight that helps to capture opportunity." It provides a new vantage point for conceiving a solution that is not only within reach, but may also, in hindsight, even seem obvious. Throughout this book, we'll describe many real-life examples of how aha moments can reveal unforeseen simplicity.

EUREKANOMICS

Economists and business leaders increasingly emphasize creativity and innovation as a spur to economic growth. Globalization and technological advancement have, as Pulitzer Prize–winning *New York Times* columnist Thomas Friedman phrased it, "flattened the world" by diminishing the developed nations' historical advantages of geography and wealth. Computers are now relatively inexpensive. The Internet makes it just as easy to tap the brains of Beijing as it is to tap the brains of Boston. Consequently, the people and businesses with the best new ideas—no matter who or where they are—increasingly have the best chance of prospering. These ideas frequently are the products of creative insight.

Recognizing the importance of creativity to economic success and incited by proclamations of a burgeoning "innovation gap," governments, corporations, and private institutes have been considering policy reforms to help the development of cutting-edge products, methods, and efficiencies. These efforts are based on the assumption that the main impediments to innovation are defective organizational structures that block the implementation of original ideas. The hope is that creativity could be liberated if governments and corporations would simply change their policies.

Though there is some truth to this view, it doesn't tell the whole story. Implementation isn't the only bottleneck—in fact, Insightfuls sometimes find creative ways to circumvent such difficulties. Some of the biggest impediments, perhaps the most stubborn ones, are mental rather than institutional. Groups of people can inspire, refine, and implement ideas. Organizations can support and foster creativity. However, innovations can't be implemented if they aren't conceived in the first place. The basic truth is that ideas originate in individual brains. Individual creativity must therefore be thought of as a precious resource to be sought and cultivated—especially when there is reason to believe that it may be dwindling.

HOW SMART ARE WE?

Intelligence tests have been around for about a century. These tests measure analytical skills involved in logical reasoning rather than creativity. The general principle underlying intelligence quotient (IQ) tests is that a person's raw test score is compared to the raw test scores of peers. For example, a child's score is compared to the scores of other children of the same age; an adult's score is compared to the scores of other adults. An IQ score of 100 is defined to be average; half of test takers score higher than 100, while half score lower. Every few years, the scoring process is calibrated, or "renormed," to adjust for fluctuations over time. For example, the average raw score in one decade could be 70 percent of the problems solved correctly, 75 percent in the next decade, and 65 percent in the decade after that. In all these cases, the average raw score is assigned an IQ of 100 for that period of time. So, even though people's average raw scores may fluctuate, the average IQ never changes. It's always 100.

About thirty years ago, New Zealand intelligence researcher James Flynn started analyzing changes over time in these raw scores. What he found—now known as the "Flynn effect"—startled the world: Average raw scores have been *increasing*. The worldwide "new normal" has been getting better and better. We don't yet know for sure why this is happening, but the numbers don't lie. People are getting smarter. *Aren't they?*

Well, sort of.

In 1966, psychologist E. Paul Torrance created the Torrance Tests of Creative Thinking (TTCT), a widely used instrument for measuring creative thinking ability rather than the kind of analytical thought measured by IQ tests. Like any other proper psychometric test, it, too, is periodically calibrated so that an individual person's score can be compared to his or her contemporaries' scores. Psychologist Kyung-Hee Kim of the College of William and Mary has recently examined changes over time in TTCT raw scores in the United States and concluded that, on average, people seem to be becoming *less* creative over time. With all our educational and technological advancements, and in spite of our Flynn-effect increases in analytical intelligence, as a society, Americans are apparently becoming less creative. Future research will show whether this drop in creativity scores is global and mirrors the worldwide increase in analytical intelligence.

This apparent waning of creativity is alarming because it comes just at a time when creative insight is desperately needed to parse our problems and reveal our opportunities. Simply documenting this deterioration doesn't explain it. By showing how eurekas happen and how they are an empowering factor in your life, we will also show how the decline can be explained—and reversed.

2

INSIGHT ILLUSTRATED

The tune itself came just complete . . . so you've got to believe in magic.

—Sir Paul McCartney, musician

Insights can be about anything and can occur in so many different contexts that we often think about the phenomenon in inconsistent ways. Let's sharpen the picture with a guided tour of some of insight's most distinctive features. This will set the stage for understanding its inner workings in the chapters that follow.

STEPWISE

Insight is one step in a sequence (see figure 2.1). In the first step, you grapple with a problem. This is sometimes called "immersion" in all of its aspects: the facts, the tools at your disposal, and your goal. You may consciously try to solve the problem, or you may just study it. If you do try to solve it and get stuck, you've reached an "impasse" and have no idea how to proceed. Or perhaps your efforts were just inter-

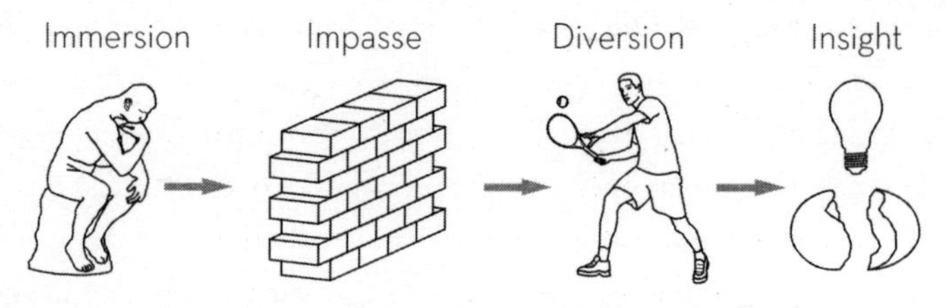

FIGURE 2.1: Stages of thought leading to insight.

rupted by something else. Either way, there is a "diversion," or a break from the problem. Then, at some point, the diversion is itself interrupted by an aha moment that interjects the solution. This is insight, also called the stage of "illumination."

This is the classic way to think about insight, although there are several elaborations and variants of this basic scheme that we will introduce at appropriate points. Now let's consider some concrete examples that illustrate the steps and features of the insight experience itself—in particular, their expanded perspective, sudden occurrence, reinterpretation of the familiar, awareness of unforeseen relationships, subjective certainty, and emotional thrill.

VANTAGE POINT

Cancer posed a deep mystery for surgeons. When a tumor is removed, tiny metastatic tumors in other parts of the body can grow quickly, with dire consequences. Paradoxically, even shrinking the primary tumor with radiation or chemotherapy can sometimes cause small secondary tumors to grow. This was one of the great puzzles of cancer research, noted in the medical literature as early as 1895. It has posed a dilemma for doctors treating cancer patients: whether to treat the primary tumor in the most aggressive manner possible.

Dr. Judah Folkman, a Harvard Medical School graduate, was doing his surgical residency at Massachusetts General Hospital when he was drafted into the U.S. Navy in 1960. The navy had just built the USS *Enterprise,* a nuclear-powered aircraft carrier with a crew of nearly four thousand. One of its advanced features was that it could stay at sea for a year at a time without having to return to port for refueling or restocking. In practice, though, this capability remained out of reach, because its operating rooms needed to keep a substantial supply of fresh blood on hand. Blood has a shelf life of days, not months. To realize the full potential of ships such as *Enterprise,* the navy sought to develop a blood substitute with a longer shelf life. Judah Folkman and a colleague, Dr. Fred Becker, were assigned this task.

Folkman and Becker thought of using hemoglobin, the oxygen-carrying component of blood. They were interested in hemoglobin because it could be stored for long periods of time in powder form and later reconstituted by adding water. So they conducted an experiment to determine whether living tissue could be sustained and repair itself while bathed in hemoglobin. For this, Folkman and Becker used a type of tissue capable of robust growth: cancer cells.

This experiment worked, at first. The cancer cells stayed alive and started multiplying. But then something unexpected happened. These cells stopped reproducing when they formed a tumor that was about the size of a pinhead. Folkman, curious about these stunted tumors, examined them more closely. He noticed something strange. They were devoid of blood vessels.

Folkman's chance observation gripped him and sparked his first important insight: Tumors, like healthy tissues, need a substantial supply of blood to grow. Without new blood vessels to feed them, they can't grow larger than a pinhead. Folkman's visionary realization was that if he could isolate the biochemical factors that regulate the growth of blood vessels, then tumors could be controlled or killed by depriving them of blood and starving them. This was a radical change of perspective that enabled him to see possibilities that others couldn't imagine.

For many years, most researchers wouldn't accept the validity or importance of Folkman's idea that tumor growth depends on blood-vessel growth. Even those who thought that his idea could be technically correct nevertheless believed that it was unimportant because the best way to attack a tumor had to be the most obvious one—a direct assault. Poison it through chemotherapy or simply cut it out. Why bother using the indirect approach of trying to slowly starve a tumor when one could attack it in a more straightforward fashion? It took members of the scientific establishment years to grasp the value of this new perspective because they were blinded by old ideas.

Many years later, another insight suddenly added a new dimension to Folkman's thinking.

AT THAT MOMENT

After his tour of duty in the navy, Folkman returned to Massachusetts General Hospital to complete his surgical residency, after which he obtained a faculty position at Harvard Medical School. During this period, he struggled to sustain funding for his tumor research in the face of criticism and derision from establishment colleagues who were blind to the implications of his idea. His determined efforts throughout the 1970s led to incremental progress in isolating various chemicals that either promote or inhibit the growth of blood vessels that nourish tumors; nevertheless, he was still unable to answer the problem of why the removal of a primary tumor would trigger the unrestrained growth of tiny metastases. He listed this issue on a whiteboard in his laboratory as one of the "burning questions" that he wanted his team to think about.

Noel Bouck's feet hurt. She was wearing new shoes when she attended a cancer research conference in 1985. All she could think of was finding a place to sit and rest. She found a haven in the nearest conference room, where nothing seemed to be going on. But soon the room started filling up with people who had come to hear the next speaker. Bouck was quickly boxed into her row of chairs by the swelling audience and couldn't escape without a fuss, so she stayed for the talk. The speaker was Judah Folkman. His talk was about cancer and angiogenesis, the growth of blood vessels.

Bouck didn't know much about angiogenesis and hadn't thought about its relationship to cancer. Her research specialty was the genetic basis of cancer. Folkman's presentation was a revelation to her. By the end of his talk, she said to herself, "This guy is right. He's absolutely right. I believe it."

She started researching angiogenesis and cancer.

In 1987, Folkman read a new research report. It came from Bouck's lab. In this groundbreaking study, she and her collaborators demonstrated that tumors release both chemicals that cause new blood vessels to grow toward them and chemicals that inhibit blood-vessel growth. Bouck's report became required reading for Folkman's laboratory staff, and Folkman pondered its meaning daily.

In September 1989, Folkman, the son of a rabbi, went to a Rosh Hashanah service in Boston's Temple Israel. At ten a.m., sitting in the back row while listening to the Jewish New Year prayers, Folkman was struck by a sudden insight that "explained everything." The growth of blood vessels toward a tumor is controlled by a balance between the chemicals that the tumor releases to spur vessel growth and those it releases to inhibit it. When the balance is tipped in the direction of the growth-promoting chemicals, the blood vessels connect to the tumor, enabling the tumor to grow. When the balance favors the inhibitory chemicals, blood vessels do not reach the tumor. In this case, the tumor cannot grow larger than about a millimeter.

This eureka moment explained why the removal of a primary tumor can spark the growth of tiny metastases elsewhere in the body: The primary tumor releases both vessel-promoting and vessel-inhibiting chemicals. The vessel-promoting chemicals last only a few minutes in the bloodstream—not long enough to stimulate blood-vessel growth to the smaller tumors. The vessel-inhibiting chemicals last indefinitely in the bloodstream, allowing them to dominate and prevent new blood vessels from feeding the little tumors. When the big primary tumor is removed, so are the inhibitory chemicals that restrain blood-vessel growth. This is like pouring gasoline on smoldering embers.

Note the abruptness of Folkman's aha moment—so abrupt that he was later able to identify the specific time and place of its occurrence. It was like a door that is jammed shut and resists all force until it suddenly gives way and opens. And once the door is opened, everything can be seen.

Folkman's insights offered a radical change of viewpoint that was difficult for his colleagues to accept. After all, it's hard to reinterpret that which one takes for granted, even when one's life depends on it.

THE FLIP SIDE

On August 5, 1949, a team of fifteen firefighters led by foreman Wag Dodge was airlifted to Mann Gulch in Montana to extinguish what the men thought would be a relatively small brush fire on one side of the gulch. They parachuted onto the opposite side of the gulch, joined up with one fire guard, and started descending with the wind at their backs. Then, unexpectedly, the wind reversed, and the fire jumped over to ignite the grass on their side. The flames rapidly approached the men. They began to climb up the slope to try to outrun the fire. They paused briefly to drop their cumbersome equipment. But Dodge realized that this wouldn't save them. The fire was moving too quickly.

Dodge stopped. When the other firefighters saw this, they must have thought that he was giving up. They desperately continued their flight. But if Dodge did give up, it wasn't for long, because he had a lifesaving insight. With his back to the flames, he took out a match and lit the grass in front of him. The dry grass caught fire immediately, and the wind blew this new fire up the side of the gulch and away from him, leaving a patch of charred ground. Dodge then crawled onto this patch and waited. When the brush fire arrived, it flowed around—and then away from—the burned ground that sheltered him.

Dodge survived. Thirteen of the other fifteen firefighters perished.

Fire was the problem. Given this obvious fact, it wasn't easy for the firefighters to mentally flip their interpretation of the situation and see fire as the solution. Dodge's "escape fire," though familiar to the Plains Indians, was unknown to the forest service at that time and was not a part of the firefighters' training. This is why the other members of the team must have thought that Dodge had lost his judgment or was just succumbing. They couldn't imagine any solution other than what seemed to be the most obvious one: Flee from the flames. Dodge's insight was a sudden flip of his understanding. His radical reinterpretation was utterly nonobvious: Fire wasn't just the problem—it was also the *solution*. He fought fire with fire.

The conceptual restructuring that produces an insight can be relatively simple—in Dodge's case, a straightforward inversion—or it can be a lens for viewing subtler relationships.

BY WAY OF ANALOGY

Andrew Stanton of Pixar Animation Studios already had a string of hit movies to his credit. He had written blockbusters such as *Toy Story; Monsters, Inc.; Finding Nemo;* and *Cars* and was writing a new film that would prove just as successful: *WALL-E,* about the last robot left on a hopelessly polluted earth abandoned by humans many years before. One of the problems with which he had been struggling was the design of WALL-E's face. It had to be machinelike, yet expressive.

One day, Stanton went to a baseball game. He couldn't see the game very well, because his seat was "crappy" (for which he blamed his editor). So he borrowed a pair of binoculars from the person sitting next to him. He mistakenly turned the binoculars around. With the lenses on the wrong side staring at him, the answer to the problem "dropped in my lap." The binoculars looked like a face. He flexed the inner hinge a few times to create different facial expressions and saw "an entire character with a soul in it." It was settled. The robot WALL-E would look like a "binocular on a stem."

Stanton's insight caused him to miss a whole inning of the game. Undoubtedly, winning an Oscar made up for that.

This aha moment conferred an analogical insight. Analogical thinking solves a problem by revealing a deep relationship between two things that appear very different from each other on the surface. Insights aren't the only way to experience an analogy. You can deliberately try to construct one. But when you spontaneously realize that one situation is similar to another, you've had an analogical insight.

SUBTERRANEAN RUMBLINGS, SPONTANEOUS ERUPTIONS

Insights often interrupt ongoing thought, most dramatically when a person isn't even thinking about the problem whose solution sud-

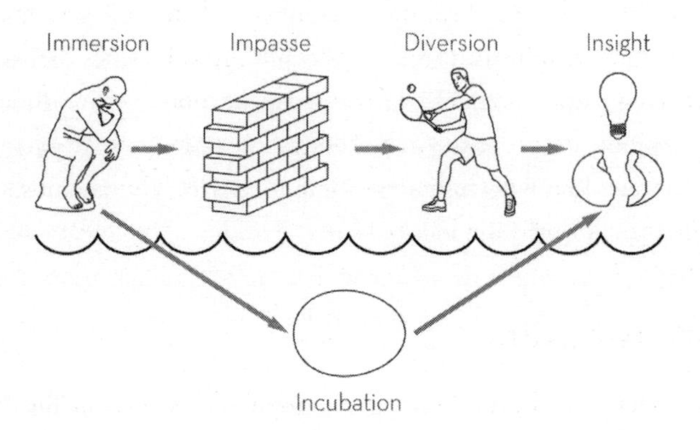

FIGURE 2.2: Insights are generated by unconscious incubation.

denly presents itself. Though abrupt, like volcanic eruptions, insights are culminations of sustained underground activity called "incubation."

Judah Folkman hadn't been thinking about cancer when he realized that tumors couldn't grow beyond the size of a pinhead without recruiting a blood supply. However, seeing tiny tumors devoid of blood vessels ignited an incubation process that sparked his idea that controlling the growth of blood vessels would allow one to control the growth of tumors. Later, he spent many years thinking about cancer and angiogenesis, which put all the elements of his next insight into place. These pieces eventually snapped together when he ducked out of his lab to attend a religious service. Similarly, Andrew Stanton had been struggling for some time with the problem of a face for WALL-E. The answer appeared when he took a break from work to enjoy a baseball game.

Such examples show that an insight can incubate unconsciously in your brain while you're thinking about other things, only to pop into consciousness at an unexpected moment, sometimes triggered by a seemingly irrelevant stimulus such as bloodless tumors or a pair of binoculars. Later we'll see how an insight's path to awareness can also be greased by a change of setting or context and an improved mood. (Wag Dodge's insight during a crisis is not typical, as we discuss in the notes to Chapter 9.) A change in setting or mood neither directly causes insights nor prevents them, though they can promote particular brain states that will facilitate or inhibit insight. These brain states are fertilizers for seedling ideas.

SELF-EVIDENT TRUTH

Barbara McClintock, trailblazing geneticist and winner of the 1983 Nobel Prize in Physiology or Medicine, is famous for many breakthroughs, such as the discovery of transposons, sections of DNA that can move to other positions within the genome of a cell. Many of her ideas leapt so far ahead of contemporary research in cytogenetics that her colleagues often didn't understand or accept her theories. Like Judah Folkman, she was sometimes ignored or spurned for these ideas. Tired of explaining and justifying her research to her less-imaginative colleagues, she stopped publishing her findings in mid-career.

McClintock, a true Insightful, had a number of aha moments that propelled her research forward, and she left vivid descriptions of some of them. In one case, McClintock was supervising a postdoctoral researcher who was studying the genetics of corn. He was examining "translocations," exchanges of genetic material between different types of chromosomes. He expected that half of his plants' pollen would be normal and half would be sterile and "heterozygous"—that is, having two different alleles for the same gene, ordinarily one of which is dominant and the other recessive. The postdoc was surprised to discover that his prediction was wrong—the plants were 25 to 30 percent sterile rather than the expected 50 percent. McClintock was perplexed by this unexpected finding and walked from the cornfield to her laboratory. She spent about half an hour considering it when "suddenly I jumped up and ran down to the field. At the top of the field . . . I shouted, 'Eureka, I have it! I have the answer!'" McClintock realized that the unexpected outcome of the experiment resulted from an extra, third, copy of the gene, an abnormal condition known as "trisomy." (Genes usually come in pairs.)

After she explained her idea, the other geneticists demanded that she prove it. However, McClintock couldn't immediately describe how she had solved the problem. "It had all been done fast; the answer came, and I'd run. Now I worked it out step by step—it was an intricate series of steps—and I came out with what it was." Intrigued by her ability to solve the problem without knowing how she had solved it, McClintock mused, "Now why did I know, without having done a thing on paper? Why was I so sure that I could tell them with such excitement and just say, 'Eureka, I solved it!'?" This sort of thing happened to her frequently. On another occasion, she said, "Something happens that you have the answer—before you are able to put it into words. It is all done subconsciously. This has happened too many times to me, and I know when to take it seriously. I'm so absolutely sure. I don't talk about it, I don't have to tell anybody about it. I'm just *sure* this is it."

CERTAINLY JOYFUL

In addition to the lack of conscious awareness of the mental processes that led to McClintock's insight, note the emotional rush that incited her to shout and run back to the other geneticists to tell them her solution. This experience of joy and importance contributes to the feeling of certainty about an insight even before the answer can be validated. This doesn't usually happen when a solution is derived analytically. You don't often cheer after adding up a column of numbers—unless the total is unexpectedly in your favor.

McClintock's emotional outburst echoed the most famous "eureka" of all, a legendary one that took place more than two millennia earlier. As the story goes, in the second century B.C., King Hiero II of Syracuse summoned the great Greek mathematician Archimedes and commanded him to solve a problem. The king had commissioned a new crown from a goldsmith but was uncertain whether the crown had been made with the pure gold for which he had paid. The mathematician's task was to determine, without damaging the crown, whether it was indeed made of pure gold. The density of gold was known. If Archimedes could calculate the volume of the crown, then the density of the crown could also be calculated and compared with the density of gold. If these two densities differed, then the crown couldn't have been made from pure gold. Of course, the crown could be weighed easily enough, which is one step toward calculating density. The other step involves calculating the object's volume. However, given its irregular shape, there was no clear way to determine its volume without melting it down and recasting it into a regular shape, such as a cube, for which there is a simple method for computing volume.

Both the weight and the volume of the crown were necessary to calculate its density, but Archimedes had only the weight, so he was stuck. He couldn't use conventional mathematical methods for deriving the solution. But he wasn't the type of person to give up easily, so he continued to ponder the problem.

One day he filled a tub with water to take a bath. As he stepped into the water, something remarkable happened. He watched the water level rise as his body displaced the fluid. Then, he instantly realized the solution to the problem. According to the story, he jumped up and, without donning his clothes, ran out of the house and down the street yelling *"Eureka!"* (Greek for "I have found it!").

His sudden insight was that an irregularly shaped object placed into a liquid displaces a volume of liquid equal to that of the immersed object. It's a simple matter to measure the volume of the runoff water by placing it into a measuring container. Then, with the volume and weight of the crown known, he could compute the crown's density, which could be compared with the known density of pure gold. This newfound method showed Archimedes that the crown was not pure gold.

Like Barbara McClintock's example, the story of Archimedes illustrates the emotional rush of insight and the contribution it makes to the Insightful's certainty. This certainty is clear from what Archimedes yelled—"I have found it!" rather than "I might have it!"

However, certainty isn't the only consequence of the thrill of revelation. Like any other pleasurable experience, insights can be addicting.

MORE, PLEASE

Everyone wants to have more insights to help us solve practical personal and professional problems and to understand our world and ourselves. But we also crave the experience: the exciting expansion of consciousness, the explosion of knowledge, the burst of joy, the certitude. As Carl Sagan said, "When we think well, we feel good. Understanding is a kind of ecstasy." That's why people read mystery novels, tackle crossword puzzles, and seek answers to the big questions of existence. And that's why, not just for profit but also for pleasure, many people try to develop personal strategies for cultivating insight.

Interest in creativity enhancement is not a new development. In 1891, a seventieth-birthday dinner was held to honor the great German scientist Hermann von Helmholtz, a true Insightful whose many contributions include the principle of conservation of energy, the measurement of the speed of nerve impulses, and the invention of the ophthalmoscope that ophthalmologists use to examine the inside of the eye. In a speech he gave to the attendees of the dinner, Helmholtz offered one of the earliest accounts on record of personal strategies for achieving insights:

> Often . . . [ideas] arrived suddenly, without any effort on my part, like an inspiration. . . . They never came to a fatigued brain and never at the writing desk. It was always necessary, first of all, that I should have turned my problem over on all sides to such an extent that I had all its angles and complexities "in my head." . . . Then . . . there must come an hour of complete physical freshness and quiet well-being, before the good ideas arrived. Often they were there in the morning when I first awoke. . . . But they liked especially to make their appearance while I was taking an easy walk over wooded hills in sunny weather.

Although Helmholtz's remarks predate the earliest scientific studies of insight, his simple observations hold up very well. Note his description of immersion as turning "my problem over on all sides to such an extent that I had all its angles and complexities 'in my head.'" He also pointed out that diversion from a problem by sleep or a stroll through wooded hills can allow an idea to emerge suddenly. In fact, until a few years ago, little could be added to his prescriptions.

Recent research expands on these ideas and the classic insight framework to contribute entirely new dimensions to insight enhancement, most of which Helmholtz could have scarcely imagined. These dimensions are based on a new understanding of how the brain's design compensates for its practical limitations.

3

THE BOX

The difficulty lies, not in the new ideas, but in escaping from the old ones, which ramify, for those brought up as most of us have been, into every corner of our minds.

—John Maynard Keynes, preface to *The General Theory of Employment, Interest, and Money* (1935)

The sixteenth-century historian Girolamo Benzoni told a story about a dinner party that Christopher Columbus attended with a group of Spanish nobles. It would have been a brilliant affair—the aristocracy of the day entertaining the man who discovered the New World. The nobles wanted to hear stories of his expeditions and share in the glory by association with the great man.

But not all of them were impressed. One commented that Columbus's discovery of a new route to the Indies was not so remarkable, because anyone with a fleet could have done the same thing. Columbus then asked for an egg. He challenged the nobles to make the egg stand on its end. They each tried and failed. Then Columbus tapped the end of the egg gently on the table to make a slight indentation in its shell—we can safely assume that it was hard-boiled—and stood it up on its flattened end. The nobles immediately understood: The solution to a problem may seem obvious after the fact, but that doesn't mean that it was obvious beforehand. This is what's meant by the phrase "Columbus's egg." It's an idea that is obvious only in hindsight (see figure 3.1).

Note the expression on the face of the nobleman standing behind Columbus and looking over his shoulder: It expresses an "I should have known that!" moment.

The Columbus story eventually became attached to the "Christopher Columbus Egg Puzzle" depicted in Sam Loyd's 1914 *Cyclopedia of Puzzles* (see figure 3.2).

FIGURE 3.1: William Hogarth's illustration of Christopher Columbus standing an egg on its end. The expression on the face of the nobleman standing behind Columbus expresses an "I should have known that!" moment. *Wikicommons (en.wikipedia.org/wiki/File:Columbus_egg.jpg)*

FIGURE 3.2: The Christopher Columbus Egg Puzzle. *Loyd, S. (1914), Cyclopedia of Puzzles (en.wikibooks.org/wiki/Creativity_-_An_Overview/Thinking_outside_the_box#mediaviewer/File:Eggpuzzle.jpg)*

This puzzle evolved into one that is familiar to most people who have taken an undergraduate cognitive psychology course. Called the Nine-Dot Problem, it was the focus of a famous 1930 experiment by psychologist N. R. F. Maier and many other studies since then. Look at figure 3.3. Your task is to connect all the dots by drawing no more than four straight lines without lifting your pencil from the paper or retracing any lines.

This problem has more than one solution. Nevertheless, most

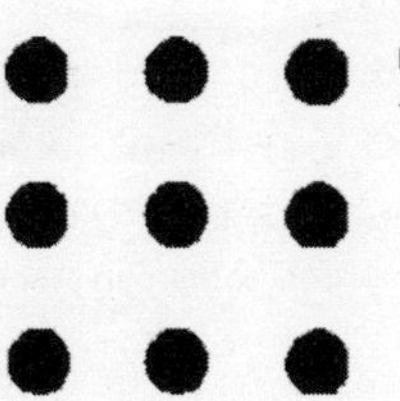

FIGURE 3.3: The Nine-Dot Problem.

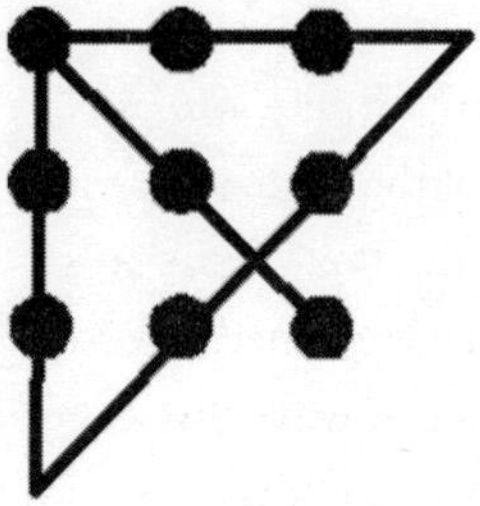

FIGURE 3.4: A solution to the Nine-Dot Problem.

people are at least initially stumped by it. In fact, it's hard to find someone who can solve it within ten minutes. The best-known solution is depicted in figure 3.4.

There is another striking solution, shown in the notes for this chapter. Give it a shot. Hint: To help you to think of the second solution, remember that this is called the Nine-Dot Problem, not the Nine-Point Problem.

Try this problem on a friend and let him or her work on it for a while. If the friend doesn't solve it within a few minutes, then reveal either solution. It's likely to elicit a groan, because the answer is so simple. The few who are able to solve it typically experience an aha.

The reason that most people find the Nine-Dot Problem so difficult to solve is that they assume that the lines cannot go beyond the boundaries of the square formed by the outer dots. Of course, this problem can't be solved without extending lines beyond the square, and the rules don't forbid drawing such lines. However, as the Gestalt psychologists showed about a century ago, people naturally tend to see a figure like the nine-dot square as a single unified object against a background rather than as a collection of nine separate objects. Reading is a familiar example of this principle. When you read words on a page, you tend to think of the text as being an object. You don't pay attention to the margins, and you certainly don't think of the margins as relevant to the text. Similarly, a would-be solver of the Nine-Dot Problem assumes that the background is not a part of the problem, so she is lured into thinking of the square as delimiting the area in which she can draw. To come up with the solution, she must literally think "outside the box." This is commonly held to be the origin of the famous phrase.

Thus, "the box" came to represent the knowledge and assumptions you have gleaned from experience. It constrains your thinking and behavior in ways that can cause you to miss a better idea or perspective. Insight allows you to transcend the box's unnecessary limitations and assumptions and grants you the freedom to consider all of the possibilities of a situation. That's what it means to think outside the box.

Researchers have been using Columbus's egg sorts of problems for nearly a century to study insight. N. R. F. Maier performed one of the most celebrated of these studies in 1931. The puzzle that he posed to people is known as the Two-String Problem (see figure 3.5). A person is led into a room in which two strings are hanging from the ceiling. Also present are a chair, a pair of pliers, and other common objects. The person's task is to tie together the ends of the two hanging strings. This is difficult because the strings are hanging too far apart for anyone to grab both ends at the same time.

Only 39 percent of Maier's participants were able to find the solu-

FIGURE 3.5: The Two-String Problem.

tion in ten minutes without a hint. The hint given to those who couldn't solve it was that the experimenter would "accidentally" brush against one of the strings, causing it to swing slightly. This helped them to discover the solution: Take the pair of pliers, tie it to the end of one string, and swing the string like a pendulum. Then, grab this string on the upswing while holding the other string and tie the two strings together.

With or without a hint, when a person finally solves this problem, she usually has an "Aha!" experience. For this reason, the Two-String Problem, like the Nine-Dot Problem, has come to be thought of by cognitive psychologists as an "insight problem" rather than a puzzle that is usually solved analytically.

What makes this simple problem so difficult for most people is that, after years of experience thinking about pliers as having one particular function, it's difficult to think of them as having a different function. This "functional fixedness" is similar to the difficulty of the Nine-Dot Problem: People assume that there are rules, boundaries, or restrictions where there aren't any. It's what the firefighters on Wag Dodge's team experienced when they couldn't think of fire as anything other than a threat to be eliminated or avoided. Only Dodge was able to overcome functional fixedness and realize that fire was also a tool that would enable him to solve the problem and save his life. In such cases, the trick is to be open to alternative, nonobvious interpretations. This is how he broke out of his box.

Functional fixedness imprisons thought just when it needs to be liberated. And, worse, it doesn't take a lifetime of experience with fire or tools for this to happen.

In 1951, psychologists Herbert Birch and Herbert Rabinowitz expanded on Maier's two-string study by first providing different experiences to their participants. Some were told to install a switch into an electrical circuit board. Others were told to install another type of electrical component called a relay into a circuit board. Then, all attempted the Two-String Problem. The only change was that, instead of a pair of pliers, the items present in the room included a switch and a relay.

Remarkably, all those who had previously installed the switch into the circuit board chose to tie the relay to the string to turn it into a swinging pendulum; and almost all of the participants who had installed the relay into the circuit board tied the switch to the string. So even a few minutes of experience installing an electrical part into a circuit board made it virtually impossible for a person to think of that same part as a pendulum weight. (A third group of participants in a control group weren't given any task to perform but had prior experience with relays and switches; exactly half used the switch, and half used the relay as a weight.)

Now it gets weird. The participants were questioned after the experiment, and their justifications for their selections were bizarre. Those in *both* the switch-pendulum group and the relay-pendulum group told the experimenter that they had selected their item to be a pendulum weight because it was more compact, heavier, or easier to attach to the string. And many of them became defensive when asked to explain their choices. Some belligerently told the experimenter, "Any *fool* can see that this one is better." They didn't have a clue about what was really going on.

Birch and Rabinowitz's study shows that a very brief experience can lead to the quick construction of a mental box that limits thought. Just as important, people are blind to the fact that they are in a box at all. The implications of these findings are staggering and cannot be overstated: Everything you do has the potential to limit what you do next. Everything you think has the potential to limit what you think next. And you won't even know that this is happening.

The fact is that you are the sum total of your experiences, even fleeting ones. Your past informs your assumptions, beliefs, and expectations but also limits your ability to think and act flexibly. As we'll see, this core principle of the human brain has ironic repercussions. First, some background.

WHAT THE CORTEX DOES

The early cognitive psychologists were working in relative ignorance of the brain. They didn't have tools that would have allowed them to observe it at work. Before the development of these methods, neuroscience was, with a few notable exceptions, focused more on how the brain is put together rather than how it operates to enable a person to think. Modern technology now allows us to measure or visualize ongoing neural activity. From this research has emerged a revolutionary new understanding of how the brain works, especially its cerebral cortex, the outer layer of neurons responsible for higher intellectual abilities.

Consider human society as an analogy for the brain. Society consists of billions of people interacting with one another in a multitude of ways. Each of us influences the people with whom we come in contact. Sometimes, we facilitate the actions of others. When you buy something in a store, your money enables the owner to pay salaries to her employees, empowering them to do things that they want to do. Conversely, a person might inhibit the actions of others, such as when the presence of a security guard discourages a potential shoplifter from stealing a desirable item. All of these individual influences combine to create activity at the level of society. Various groups emerge, such as nations, ethnicities, religions, political parties, tribes, and corporations. The activity of people within each group is coordinated or synchronized so that they can work together to accomplish various goals. For example, some groups of people end up working together to make, sell, or buy products. Others collaborate to develop and propagate ideas. Yet other groups heal people or make war. From the apparent chaos of billions of people functioning independently arise relative order and purpose. Brains work this way, too.

A typical brain contains about eighty-six billion nerve cells called neurons, mostly microscopic, which process information. (By comparison, there are only about seven billion people in the world.) Each neuron excites or inhibits the activity of hundreds or thousands of other neurons. From these complex interactions emerge groups of neurons in the cortex whose activity is synchronized, each group performing specialized functions related to seeing, hearing, thinking, feeling, remembering, and moving. But these groups of neurons don't just passively react to the world. They are tuned to expect the unexpected. This is most clearly revealed by a familiar method for measuring brain activity.

Because neurons function in synchronized groups, their individual activity combines to make electrical fields strong enough to (safely and noninvasively) measure with electrodes placed on a person's scalp (see figure 3.6). This is the electroencephalogram, or EEG. It measures electrical waves generated by the cerebral cortex.

An EEG record shows changes in electrical brain activity from millisecond to millisecond (see figure 3.7). We can analyze this recording to observe how neural activity changes over time and how

FIGURE 3.6: EEG measured with electrodes embedded in an elastic cap. The words on the computer monitor are a type of puzzle that we describe in Chapter 4. *John Kounios and Mark Beeman*

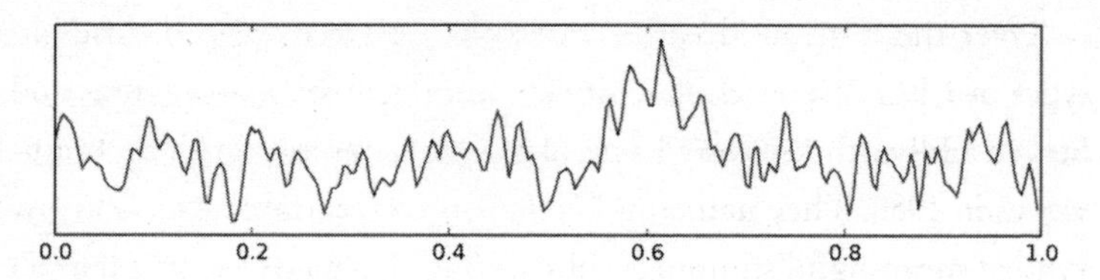

FIGURE 3.7: One second of EEG from one electrode. Time goes from left to right. Upward and downward deflections represent voltage changes over time. *Wikicommons (en.wikipedia.org/wiki/File:Eeg_raw.svg)*

the cortex responds to events as they occur. EEG doesn't give precise information about *where* something is happening in the brain. But EEG does give precise information about *when* a particular brain event is occurring. This makes it useful for detecting quick neural responses, such as surprises and insights.

BUILT FOR SURPRISE

Imagine that you are a participant in an EEG experiment. You're sitting in front of a computer screen, watching the following sentence appear on the screen, one word at a time, with each word replacing the previous word at the center of the screen:

He
takes
his
coffee
with
cream
and
dog.

Your conscious reaction to this sentence is probably nothing more than mild amusement. Your brain has a different reaction.

Over the last few decades, researchers have cataloged different types of EEG responses. Cognitive neuroscientists Marta Kutas and Steven Hillyard discovered one of the most famous of these brainwaves in 1980. They named it "N400." It's a cortical response to any kind of meaningful stimulus—like a word or a picture—that doesn't fit its context.

When you read "He takes his coffee with cream and dog," each of these words sparks a small EEG response in your brain except for "dog," which elicits a large N400. If the last word of the sentence had been the expected "sugar" instead of "dog," then there would have been little or no response because "sugar" perfectly fits the context. Figure 3.8 illustrates a typical N400 brainwave from a study by cognitive neuroscientists Phillip Holcomb and John Kounios. The graph shows the brain's EEG response to the last word of sentences such as "He takes his coffee with cream and dog," or "He takes his coffee with cream and sugar." Note the large N400 blip for "dog" and the much smaller one for the expected "sugar."

The N400 is a prime example of how the brain protests when its expectations aren't met. Every time you hear a word, your brain guesses what the next word will be and rings an alarm of surprise and confusion if its prediction is wrong. But the N400 isn't the only complainer. The brain responds with other kinds of brainwaves to show its surprise in various situations. When your doorbell rings, your brain lights up. *Who could it be? I wasn't expecting a visitor!* If you work at a typical office and see a coworker wearing a Hawaiian shirt instead of the expected business suit, your brain screams for an explanation.

These kinds of EEG responses illustrate a deeply fundamental property of the brain. We continuously build mental models of the world around us—our boxes—to help predict what will happen next. Any deviation from these expectations causes a group of neurons to shout in synchrony to signal that something is different, unexpected, or just wrong. And these brainwaves aren't pointless alarms. They

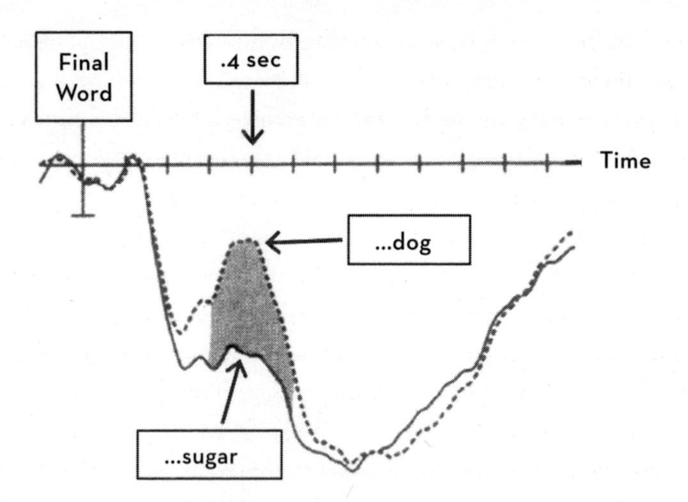

FIGURE 3.8: A typical N400 brain response, adapted from a study by Phillip Holcomb and John Kounios. Time goes from left to right on the horizontal axis. The size of the N400 is indicated by the height of the brain wave from .3 seconds to .5 seconds after the presentation of the word "sugar" or "dog." *John Kounios and Mark Beeman*

are critical to the brain's efforts to keep itself current and accurate. Things change. Like successful politicians, brains constantly test the waters and revise their policies. If a brain (or a politician) can't keep up with the latest trends, then it can't make correct choices. And whenever one of its predictions is proven wrong, this kicks off a cascade of further neural activity (or staff meetings) to figure out what happened and how the situation should be handled.

In fact, by preparing for an expected event, your brain actually sensitizes itself to unexpected happenings, should they occur. When you're driving and you stop for a red light, after a few seconds of waiting, your brain starts generating a "CNV" (contingent negative variation) brainwave that signals your rising expectation that the light is about to turn green. When that happens, you press your car's accelerator pedal but otherwise barely take notice of the green light itself. However, because you prepared yourself for a green light, if it went from red to yellow, you would be flummoxed. Without expectations, there can be no surprise.

If you step back and look at the big picture, it's clear that much of the work that a person's cerebral cortex does involves constructing, maintaining, checking, and, if necessary, modifying its mental models of the world—its boxes. These expectations are automatically and unconsciously triggered by stimuli around us, such as people, objects, situations, locations, and so forth. This enables a brain to make quick, moment-by-moment predictions and interpretations based on its experience and knowledge. Thus, the human brain is, at its core, an *anticipation machine*.

This ability, however useful, sometimes incurs a cost: tunnel vision. Each of us is condemned to live in a box. It's the way we're built. The negative consequences are usually minor, such as the inability to think of turning a screw with a dime when a screwdriver isn't handy. However, this can also prevent us from solving more important problems, such as whether or when to look for a new job. Worse yet, a maladaptive box can even force a person to interpret and respond to the world in pathological ways, as in a phobia of germs.

No one wants to be a slave to his or her knee-jerk expectations. Are we ever free? Of course, the occasional insight can provide a brief furlough from this conceptual prison. But are aha moments the only times when we can experience real freedom of thought? Perhaps the innocence of childhood holds a clue.

CHILD'S PLAY

One day, John popped into a convenience store to buy a cup of coffee. He poured his coffee and took the cup to the table that had the cup covers, sweeteners, and other coffee-related accessories. The coffee cup was very hot, so the first thing on his mind was getting a cardboard collar to place it in. His hand was starting to feel uncomfortable from the heat, but he had to wait because the store was crowded and three people were blocking his path to the cup collars. They were a mother with her two children, a young girl and a slightly older boy. The mother had a hot cup of coffee in her right hand and urgently wanted to slip it into a collar. Her left hand was carrying a couple of things she wanted to purchase, so she asked the boy to take a collar from the rack, which he did. The collar was pressed flat, and she asked him to open it and hold it on the table so that she could slip her cup into it. He didn't understand what she was asking him to do with this flat thing. She explained in more detail that he should hold it between his thumb and fingers and squeeze to open it up. He still didn't get it. After all, how could a piece of cardboard "open up"? He tried to wrap the flattened cardboard around the cup she was holding. The mother had a pained expression on her face—one that John probably shared—because her hand was getting very hot. At this point, the little girl, clearly frustrated by her big brother's inability to see what could be done with the flattened collar, grabbed it from his hand, pinched it so that it expanded into a ring, and then placed it on the table for her mother to slip the cup into. Just in the nick of time.

Generally, older children solve problems better than younger ones. However, even the short life experience of older children can sometimes be a hindrance when it's necessary to look beyond the obvious to solve a problem. In these kinds of situations, younger children can sometimes have greater freedom of thought. Developmental psychologists Tim German and Margaret Anne Defeyter showed five- , six- , and seven-year-old children a miniature room that contained a teddy bear that they were told wanted a toy out of reach on a high shelf. The room also contained a carton and a few other objects that the bear could potentially use to help solve the problem. Unsurprisingly, the older children were more likely than the younger children to use the carton as a platform for the bear to stand on and reach the toy. The few younger children who were able to think of this solution needed more time than the older ones to figure it out.

That's what happened when the carton was presented separately from the other objects. For other kids, it was initially filled with the other objects, reminding them of a carton's usual function as a container. When the carton was presented as a container, the younger children were more likely than the older ones to think of it as a platform for the bear to stand on. They also thought of this solution more quickly than the older kids. The little ones' naïveté empowered them to consider the wider possibilities that eluded the older children.

But a lack of life experience may not be the only source of a young child's conceptual freedom. Don't forget that kids aren't just undersized adults. Their brains are immature rather than just smaller. A child's brain takes many years to achieve an adult state. Among the last of the brain structures to fully mature is a part of the cerebral cortex called the "frontal lobe" (see figure 3.9), which plays a key role in the higher-order "executive functions" involved in problem solving, such as planning, setting goals, switching attention, and selecting concepts. One function of the frontal lobe is to limit the range of be-

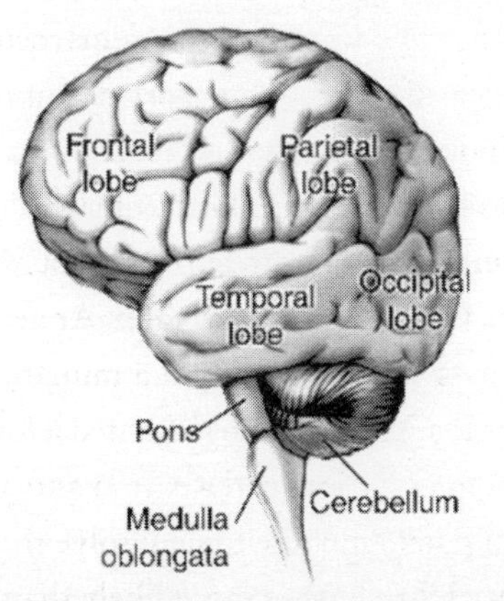

FIGURE 3.9: The major externally visible structures of the human brain. This diagram shows a view of the left side of the brain. The cerebral cortex is the outer layer of neurons comprising the four "lobes" shown in the diagram. *Wikicommons (commons.wikimedia.org/wiki/File:Brain_headBorder.jpg)*

haviors a person will consider in a given situation. For example, if you order an ice cream sundae and the server asks you what topping you would like on it, you'll probably consider things such as chocolate syrup, fruit, whipped cream, peanuts, and sprinkles. The frontal lobe is involved in narrowing the possibilities you consider to things such as these and excluding from consideration unusual, but possible, sundae toppings such as squid or curry.

OK. The frontal lobe carries out some important cognitive functions, but it's beginning to sound like it's also the jailer that keeps us trapped in the box. What would adults be able to accomplish if they were freed from its restraints? Italian neuropsychologist Carlo Reverberi and his colleagues took this logic to its extremes. They reasoned that if the frontal lobe serves to restrict one's thinking to the usual suspects, then it could actually be an impediment to insight. So they posed insight problems to patients with frontal lobe damage and to healthy control participants.

The patients did better.

If this is giving you the idea of making an appointment with a neurosurgeon, you might want to hold off. Removing your frontal lobe wouldn't help you to break out of the box at will. Instead, it would simply destroy the box. And if there is no box, then there is nothing to break out of. Reverberi's patients weren't transcending the boundaries of thought. Their thought had no boundaries to transcend. This made them better able to consider nonobvious solution possibilities. However, it's only in such rare, specific circumstances that frontal lobe patients have any kind of advantage. After brain damage, people do most things worse than before the damage. Similarly, younger children do most things worse than older children. It's not that children or frontal lobe patients are especially "creative" in the sense that we apply this term to healthy adults. Rather, they lack boundaries to guide their thinking. They can't draw on their knowledge and experience to know what kinds of interpretations and solutions are likely to be helpful in a particular situation. This lack of interpretative constraint can sometimes result in ideas that appear to be creative but are more often just irrelevant or silly.

In other situations, constraints confer distinct advantages.

THINK SMART, NOT MORE

When John was a teenager, he played a game of chess against the late Danish Grandmaster Bent Larsen, one of the strongest players of his era, in a "simultaneous exhibition" in which the Grandmaster played dozens of competitors at the same time. He watched as Larsen walked from chessboard to chessboard, briefly glanced at each game, made a move, and then walked on to the next game. After John had made a few questionable moves, he believed that Larsen would soon crush him. But when the Grandmaster arrived to make his next move, he barely glanced at the board and then . . . offered John a draw! Flabbergasted, John accepted the offer, thinking that perhaps Larsen just didn't feel like going through the tedious technical exercise of finishing off a mortally wounded amateur. Or perhaps, out of compassion, Larsen was just trying to spare a pathetic duffer from embarrassment. But neither of these possibilities made sense—chess Grandmasters are ferocious competitors who would never pass up an opportunity to defeat any opponent. Later, John analyzed his record of the game to figure out whether the offer of a draw could be explained by something other than a hurried schedule, impromptu benevolence, or sheer luck. In fact, lengthy analysis later showed that Larsen could not have gone on to win, assuming that John made no further bad moves. (This was perhaps a charitable assumption on the part of the Grandmaster.) Offering a draw was therefore the right thing for Larsen to do, or at least would have been if he had been playing against a stronger opponent.

How did a mere glance tell Larsen what took the weaker player hours of analysis to figure out? John's aha moment: The Grandmaster accomplished this by thinking *inside* the box.

People who have great knowledge and skill in some area are experts. Being an expert, such as Bent Larsen, means that you have a bigger and better box to guide you to superior performance in your specialty. To understand this benefit, it's important to understand precisely what expertise really is.

A common fallacy about experts in any domain is that they think more, or at least they think more quickly, than nonexperts. Nothing could be further from the truth. Chess experts such as Bent Larsen don't necessarily calculate more possibilities than amateur players do. José Capablanca, an early twentieth-century world chess champion who was one of the game's greatest players, was once asked how many moves ahead he could calculate. His legendary answer: "Just one, the best one."

Instead of outcomputing their opponents, Grandmasters win by being better at recognizing patterns that emerge during a game. Their vast experience and memory for games they've played and analyzed tells them which avenues are promising and which are dead ends that can be ignored so they don't have to figure out all the possible consequences of all the possible moves. This is how experts get better results while thinking less, not more.

Research supports this view of expertise in chess and beyond. Expert chess players remember meaningful chess positions better than players who aren't experts; but, surprisingly, experts don't remember random or nonsensical positions significantly better than nonexperts. This is because meaningful patterns serve as a kind of shorthand that's easier to remember than a meaningless configuration of pieces that couldn't occur in a real game. If you showed a chess position to a Grandmaster and asked him to memorize it, he may immediately recognize that the position is, with one slight exception, the same as one that occurred in a game played by Capablanca in the Nottingham tournament of 1936. Instead of remembering all the individual positions of the chess pieces, all that has to be remembered is a single label plus any small deviations from the pattern, for example, "Capablanca's 1936 Nottingham game except that the white pawn is on e4 instead of e5." But show the Grandmaster a board with a random jumble of pieces with no meaningful pattern to recognize, and she will have little or no memory advantage over the nonexpert.

This is like learning to recite from memory a poem in a foreign language that you don't know rather than memorizing one written in your native language. In a foreign language, the poem is a series of meaningless sounds that are difficult to learn because you would have to memorize each of them individually. But in your native language, you can use your knowledge to group the sounds into words and sentences and use your understanding of the poem's meaning to help you learn the specific sequence of sounds.

Here's an even simpler example: The following seemingly random sequence of letters is relatively difficult to learn:

TIMMBIIBFASUAIC

But if you reverse the order of the sequence and group the letters into threes, you have something more familiar:

CIA USA FBI IBM MIT

This sequence is much easier to learn because you can use the trick of remembering the letters by thinking of the meaning of these acronyms.

Grandmaster Larsen apparently understood how experts operate. He was one of the most successful tournament players of his time because he found a way to render his opponents' boxes useless. Chessmasters invest an enormous amount of time studying classic games to minimize the amount of raw calculation that they have to do during a tournament. To nullify the benefits of his opponents' preparation, Larsen often used strange openings and made unexpected moves. This forced his opponents to "throw out the book," or, rather, "throw out the box," and rely on mental computation instead of preparation and pattern recognition. Neutralizing the benefits of preparation frequently enabled him to triumph because his ability to calculate his way through novel positions was often superior, especially when his adversaries were rattled by his unusual moves. This strategy was extremely effective in tournaments in which he played no more than one or two games against any single individual, but it was less effective when he played a match of many games against a single player. In such matches, after a few games his strongest opponents were sometimes able to adapt to his strategy and defeat him by minimizing their preparation and trying to play with an open mind and no preconceptions.

Larsen's strategy of neutralizing the box was appropriate only for dealing with other opponents whose expertise rivaled his own. It's simpler and easier for a Grandmaster to use his superior box to crush a weaker player.

QUICK THINK

Another benefit of the box is that it allows the expert to act quickly. Force an expert chess player to play rapidly and his performance will deteriorate little. Experts don't need much time because they don't ordinarily need to compute many possibilities. They immediately know what will work and what won't. The quick wits of experts can even save lives.

On January 15, 2009, US Airways flight 1549 took off from New York's LaGuardia Airport bound for Charlotte, North Carolina. While ascending, the plane struck a flock of birds. Captain Chesley ("Sully") Sullenberger smelled burning—birds had been sucked into the engines. Then the engines went dead. Within thirty seconds, Captain Sullenberger concluded that the engines couldn't be restarted. The plane's altitude was three thousand feet and decreasing rapidly. He communicated the situation to the air traffic control tower and looked for a place to land.

Decades of training and experience were brought to bear in an instant: "I quickly determined that due to our distance from LaGuardia and the distance and altitude required to make the turn back to LaGuardia, it would be problematic reaching the runway, and trying to make a runway I couldn't quite make could well be catastrophic to everyone on board, and persons on the ground. And my next thought was to consider Teterboro," Sullenberger later said in an interview. But then he judged that Teterboro Airport was also too far away. How did he exclude all these possibilities? Sullenberger explained:

> It wasn't so much calculating as it was being acutely aware, based upon our energy state and by visually assessing the situation, of what was and what was not possible. There are several ways I used my experience to do that. I knew the altitude and airspeed were relatively low, so our total energy available was not great. I also knew we were headed away from LaGuardia, and I knew that to return to LaGuardia I would have to take into account the distance and the altitude necessary to make the turn back. In the case of Teterboro, I knew that was even farther away, even though we were headed in that direction. The short answer is, based on my experience and looking out the window, I could tell by the altitude and the descent rate that neither [airport] was a viable option.

Thus, within one minute, he concluded that only one chance remained: to try for a landing in the Hudson River. He didn't methodically reason through all the possibilities. His expertise enabled him to be "acutely aware" of both the viable options and the dead ends.

Sullenberger also knew that only a handful of pilots had ever been able to safely accomplish what he was about to attempt. How could he do it? Again, he drew on his expertise: "I needed to touch down with the wings exactly level. I needed to touch down with the nose slightly up. I needed to touch down at a descent rate that was survivable. And I needed to touch down just above our minimum flying speed but not below it. And I needed to make all these things happen simultaneously."

Three and a half minutes after the bird strike, flight 1549, with its complement of 155 passengers and crew members, landed safely in the Hudson River. All were saved. Why did Captain Sullenberger succeed where virtually all pilots had previously failed?

Sullenberger learned to fly at sixteen. When he enrolled in the U.S. Air Force Academy, he received glider training and became an instructor pilot. In the air force, he spent five years as a fighter pilot, where he became a flight leader, a training officer, and a member of the aircraft accident investigation board. After leaving the air force, he became a commercial airline pilot, where he accumulated tens of thousands of hours of flight time, not to mention additional training in flight simulators. He developed safety protocols and training courses for flight crews, started his own flight-safety consulting business, and continued his involvement in accident investigations for the air force and the National Transportation Safety Board. But Sullenberger himself had the most succinct explanation for his heroic, virtuoso performance: "One way of looking at this might be that for forty-two years, I've been making small, regular deposits in this bank of experience: education and training. And on January 15 the balance was sufficient so that I could make a very large withdrawal."

Captain Sullenberger didn't have an aha moment. He thought within the box—a magnificent box that he had painstakingly constructed over the course of forty-plus years. His massive expertise enabled him to quickly recognize the patterns around him and apply his finely honed skills. What he did wasn't really novel. He didn't create any new ideas. But he did have more relevant knowledge and experience than other pilots, which was why he, his passengers, and crew members lived to talk about flight 1549.

The feats of such experts are spectacular demonstrations of the power of the box. But the advantages that the box affords in everyday life are no less important.

FILLING IN THE GAPS

Consider the following little story: "Tom walked to the store to buy a carton of milk. He thanked the cashier and returned home." Though this vignette seems straightforward enough, it nevertheless takes vast real-life knowledge to understand it. Consider all the things that this story does *not* explicitly state. It doesn't state that the store is within a reasonable walking distance. Nor does it state that Tom actually bought milk, or that the store even had milk, or that the store ever sells milk. It doesn't state that Tom had any money, and if he had money, it doesn't state that he paid it to the cashier. And it doesn't state that he returned home by walking. And so forth. If this story were told to a visitor from another planet—a planet that doesn't have stores, or milk, or money, or feet to walk on—the alien would be confused by the gaps in the narrative. However, you, as an earthling, automatically and effortlessly filled in the gaps while you were reading and assumed that the information had been there all along. Your expertise in ordinary life—your box—empowered you to do this.

But there's more.

STAYING AHEAD OF THE CURVE

Right now, you're reading this book. You might be sitting in an easy chair in your den with a snack or a drink on a table next to you. Perhaps you're on the train reading during your morning commute. Or perhaps you're playing out some other familiar scenario, or "script," as cognitive psychologists call them.

We have scripts for everything. For example, there is a "restaurant script." Everyone knows what typically happens when you go to a restaurant. You walk in the door and go to a stand where someone greets you and either takes your name or has a server seat you immediately. When you're seated, the server brings a menu and asks what you would like to drink. The server leaves and comes back to take your order, and so forth. This all seems rather standard and even inevitable. How could an outing to a restaurant go any other way?

Well, clearly, it *could* go differently. Differences might include self seating, a verbally presented menu rather than a printed one, candles rather than electric lights, and so forth. Someone might even come up to you and drop a bag with $10,000 in your lap, or your server might be Angelina Jolie. These things aren't likely to happen, but they *could* happen. After all, *someone* wins the lottery.

By expecting what's likely to happen next, you prepare for the few most likely scenarios so that you don't have to figure things out while they're happening. It's therefore not a surprise when a restaurant server offers you a menu. When she brings you a glass with a clear fluid in it, you don't have to ask if it's water. After you eat, you don't have to figure out why you aren't hungry anymore. All these things are expected and are therefore not problems to solve.

Furthermore, imagine how taxing it would be to always consider all the possible uses for all the familiar objects with which you interact. *Should I use my hammer or my telephone to pound in that nail? Could the oven dry my clothes just as well as that clothes dryer?* On a daily basis, functional fixedness is a relief, not a curse.

That's why you shouldn't even attempt to consider all your options and possibilities. You can't. If you tried to, then you'd never get anything done.

So don't knock the box. Ironically, although it limits your thinking, it also makes you smart. It helps you to stay one step ahead of reality. But also remember that your box isn't perfect. It can't enable you to anticipate everything, because doing so would mean that your brain would have to be as complex as the world around you. For those occasions when your box is insufficient, you have insight as a way to break out of an old box and swap it for a new and improved one.

4

ALL OF A SUDDEN . . .

Over the years I have found that it is difficult if not impossible to bring to [the] consciousness of another person the nature of his tacit assumptions when, by some special experiences, I have been made aware of them. . . . One must await the right time for conceptual change.

—Nobel laureate Barbara McClintock,
letter to maize geneticist Oliver Nelson, 1973

When John Kounios was in high school, he took a year-long calculus course. He had a B average but wasn't really excelling. He could solve problems that were minor variants of the ones he had studied and practiced, yet he had trouble with nonstandard problems. He understood all the bits and pieces but saw them as a collection of isolated procedures rather than an integrated body of knowledge. It wasn't clear to him how all the parts fit together, which meant that he wasn't able to think creatively about the subject.

This was stressful, because he really wanted to get an A for the course. His last chance to do this was by getting an A on the final exam, so he invested a lot of time in studying for it. Nevertheless, the material still didn't seem to come together.

The final exam was scheduled for a sunny June afternoon. The teacher had also scheduled an optional review session for early that morning. John debated with himself about whether he should attend that session. After all the studying he had already done, of what benefit would the extra review be? Wouldn't he be better off sleeping more and going to the final exam refreshed?

Ultimately, he went to the review session. He didn't think it would make much difference, but figured that it would be good public relations to show the teacher that he was serious.

The session was more than two hours long. The teacher briskly, but systematically, summarized all the course material in a condensed outline form. She did an excellent job, but by the end of the review the calculus still seemed to John like a laundry list of unrelated techniques.

Disappointed, he left the classroom and walked down the hall. For a few seconds, he flip-flopped on the relative merits of extra reviewing versus extra sleeping. But then he realized that he had done all that he could realistically do to prepare for the exam. It would be fine. He relaxed.

That's when it happened. He had walked perhaps forty feet down the hall when he felt a jarring sensation, almost like an electric shock. In a flash, all the bits and pieces of the calculus came together. He suddenly realized how all of the techniques and concepts were interrelated. For the first time, he felt like he truly understood the math.

With a sense of serenity and confidence, John took the final exam that afternoon and got the A that he wanted. In fact, if it hadn't been for a careless arithmetic error, he would have gotten an A+. Perhaps his newfound understanding had caused him to be a bit too relaxed.

After all these years, John still sometimes thinks about his flash of insight in the hallway of West Hempstead High School. The suddenness of this experience left a vivid impression on him.

But how sudden was it, really?

People like the idea that they can just snatch a fully formed understanding out of thin air. This view is beguiling and, subjectively, it's true. Even on those occasions when we can sense that a solution or idea lurks just below the level of awareness, it's still possible to have no clue about what the idea is or how to pull it into awareness until the idea finally reveals itself. Other times, an idea may pop into a person's awareness completely unexpectedly and unbidden, seemingly coming from nowhere. Either way, from the standpoint of conscious experience, insights really are sudden.

But experience can be misleading. Perhaps the feeling of suddenness is an illusion, a misinterpretation of the emotion that accompanies an unforeseen result. If an unexpected idea comes to mind, it might feel sudden not because the idea entered awareness abruptly, but because the idea wasn't what you bargained for. Perhaps this could bestow a faux suddenness on the idea.

But does it even matter whether an idea comes to you suddenly or gradually? Either way, an idea is an idea.

True enough. But there are a couple of good reasons for figuring out whether insights really are sudden. The first one is practical. Flying blind is not efficient. It you want to have more insights, then it's important to understand where they come from and how they arrive. This knowledge will help you to chart a reliable path to enhanced creativity.

The second reason is more theoretical. Science aims for parsimony—never settle for a complicated explanation when a simple one will do the job. This is known as "Occam's razor," named after the fourteenth-century English scholar William of Occam. During the 1970s and '80s, psychologists started to amass substantial evidence for the existence of gradually changing, or "continuous," thought processes, but little support for truly sudden or "discrete" ones. In the interest of parsimony, some researchers questioned whether sudden cognitive processes exist. Insight, thought to be a quintessential sudden process, was not spared this scrutiny. When the available research was critically evaluated, there didn't seem to be much hard evidence that what we call "insight" differs from other forms of (conscious or unconscious) thought in its abruptness or any other characteristic. The only evidence for sudden mental processes was the conscious experience of the aha moment, and some researchers argued that this type of evidence is inherently unreliable and misleading. They asserted that insight is not essentially different from everyday "noncreative," or analytic, thought—creative insight is just a myth.

This was insight's moment of truth. People's subjective descriptions of their insights, when used appropriately, can be a useful form of scientific evidence. Unfortunately, they can also be misinterpreted. Objective data from carefully controlled laboratory experiments were therefore essential. During the 1990s, Roderick Smith and John devised a way to do this. But before describing their result, let's consider a more nuanced view of what might be going on under the hood when you solve a problem or get an idea. There are three possibilities, only one of which is true insight.

ELECTION NIGHT

You're staying up late to watch election coverage on television. The vote totals for each candidate are displayed on the screen as the various precincts report their tallies. It looks like candidate Thompson is beating candidate Jones, but the night is still young. Jones has a spurt when votes from his stronghold precincts are reported. But as the evening progresses, Thompson gradually expands her lead until the television network eventually calls the election. Thompson won. In this case, you didn't really favor one candidate over the other, so you didn't experience any particular emotional response when Thompson was declared the victor.

This is how analytic solving proceeds. As you work on a problem, information about the solution accumulates until you finally have the answer (see figure 4.1). This information could accrue gradually, as when water is poured into a bucket, or in little bits, as when ice cubes are dropped into the bucket. (Think of small bumps in the vote totals as each precinct reports its final tallies.) You monitor this information as it accumulates, and when you have enough, you decide that you have the final answer.

ANALYTIC THINKING

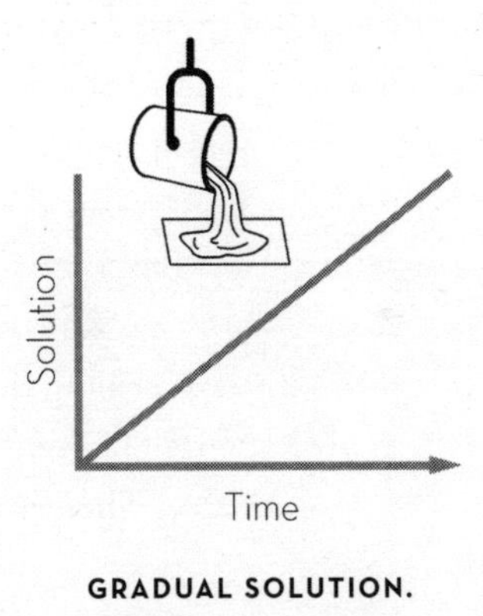

GRADUAL SOLUTION.
NO EMOTIONAL RUSH

FIGURE 4.1: When a person solves a problem analytically, information about the solution accumulates over time. A related scenario (not shown here) is one where information about the solution accumulates in little bits over time.

Now consider another scenario. You want to watch the election returns on television, but you have other things to do. Of course, the returns are being tallied all evening. You just don't know what's happening because you aren't able to tune in until late that night. When you finally turn on the TV, you suddenly discover that candidate Thompson beat candidate Jones. You are surprised and overjoyed to hear this. Thompson was your favorite.

This is how insight is supposed to work. Your brain is processing the problem, but you don't have any sense of how it's going because this processing occurs outside of your awareness. At some point, the solution pops into awareness (see figure 4.2). You are surprised and may experience an emotional rush.

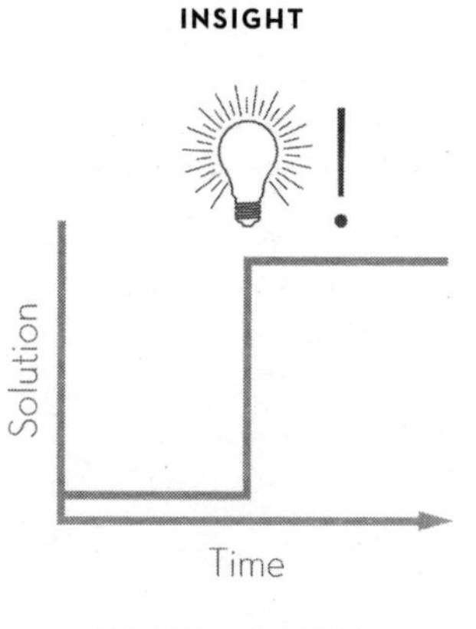

FIGURE 4.2: When a person solves a problem insightfully, he or she has no information about the correct solution until the insight arrives.

In the final scenario, you watch the election returns on TV. Candidate Thompson's lead is expanding, and you're monitoring her progress. She's getting closer and closer to victory. Then the network calls the election in her favor. You experience exhilaration because she was your strong favorite.

Let's call this "pseudo-insight." It's really analytic thought because information about the solution accumulates gradually while you watch. Nevertheless, you experience a little emotional rush when you finally solve the problem. This could be due to a feeling of exhilaration or relief when you completed your task. Or perhaps you were delighted or fascinated by the outcome. This rush made the outcome

PSEUDO-INSIGHT!

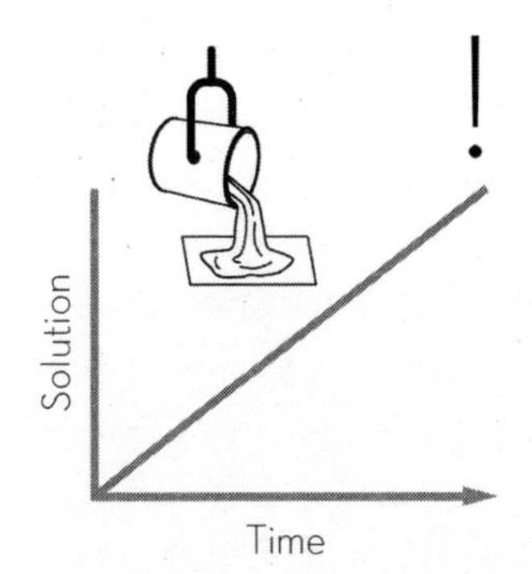

GRADUAL SOLUTION.
PLUS RUSH

FIGURE 4.3: "Pseudo-insight," in which knowledge of the solution accumulates gradually (or in little bits), but a concluding emotional rush makes the solution feel sudden.

feel like a sudden revelation even though it was really a gradual dawning (see figure 4.3).

The critical question is whether what we label as insight involves a single big jump in knowledge about the solution (figure 4.2) or whether it really amounts to "pseudo-insight"—that is, analytic thinking garnished with emotion (figure 4.3). It would certainly be easier for cognitive psychologists if they had only one type of thought to explain—gradual and analytic. Occam would undoubtedly approve of this kind of simplicity. However, if insight really were sudden, then this would mean that it's fundamentally different from analytic thought.

Of course, it's convenient when things are simple. Two kinds of thought are more complicated to explain than one type. However, it's useful to keep in mind an extension of Occam's razor that is usually attributed to Albert Einstein and is sometimes known as "Einstein's razor": "Everything should be made as simple as possible, but no simpler."

AND THE WINNER IS . . .

During the early 1990s, there was not yet a consensus among cognitive psychologists that insight was a unique mode of thought, so it was important to demonstrate that insight differs from analysis. Roderick Smith and John tackled this problem with a behavioral study. The key feature that seems to make insight distinctive is its suddenness. If we could objectively show that insights are sudden—and aren't just *experienced* as sudden—then this would distinguish insight from analysis. Our idea was to trace solutions backward in time to discover how they came into existence, whether gradually or abruptly.

We gave people a series of carefully designed anagrams to solve and used a new technique that enabled us to nudge them to guess about the solution just a fraction of a second before they would normally have solved each of these problems. We wanted to see whether these slightly faster responses showed better-than-chance accuracy. If the solution comes all at once—as a sudden insight—then *just before* the insight, the participant should have no information at all about the solution. None. For example, on election night, you would have no idea who the winner is even an instant before you turn on the TV. But if you've been watching the accumulating vote totals on TV, then a demand for a quick guess wouldn't leave you clueless. You would have a pretty good idea who the winner was.

In fact, our experiment showed that when people work on insight problems, they start with no idea what the solution is and later jump directly to the complete answer. Even an instant before they realize the solution, they have no idea at all what it is. So people can solve problems in a single bound. This means that insight is real and unique. The next step was to figure out how this happens in the brain.

5

OUTSIDE THE BOX, INSIDE THE BRAIN

A lot of it, as it is in any job in life, is being in the right place at the right time.

—Julie Harris, actress

Sometimes you get an insight just when you need one.

Jerry Weintraub managed or promoted concerts for Led Zeppelin, Neil Diamond, Frank Sinatra, and many others. He produced *Nashville, The Karate Kid, Ocean's Eleven,* and other hit movies. But as a young man in his twenties, he didn't see big success until he signed on to become Elvis Presley's concert promoter. To get this opportunity, he had to nag Elvis's manager, Colonel Tom Parker, on a daily basis for a whole year. The Colonel eventually relented.

At their first meeting, Weintraub was impressed with Elvis's gentle, unassuming personality. Unlike other big stars, Elvis didn't need a lot of special accommodations. But he was firm about one thing: He didn't want to sing to any empty seats. It always had to be a full house.

Of course, Weintraub thought that he would have no problem filling the seats, no matter how big the house. After all, this was the King.

One of Weintraub's innovations was to book Elvis, and eventually other big acts, to perform in giant arenas so that they could play to ten thousand or twenty thousand people at a time. He was able to fill New York City's Madison Square Garden and similar venues for Elvis. He also booked the convention center in Miami Beach for an evening concert. All the tickets were quickly sold out. Feeling lucky, he approached Colonel Parker about adding an extra matinee performance in the convention center; that would be an extra ten thousand tickets. The Colonel gave him the green light to arrange it.

Weintraub arrived in Miami Beach the day before the concert and went to the convention center to check on things at the box office. To his horror, he learned that five thousand matinee tickets were still unsold. It hadn't occurred to him that in July, Miami Beach is sweltering; many people wouldn't go to a daytime performance in the unair-conditioned convention center.

His professional life flashed before his eyes. By signing Elvis, he had finally made it to the big time. But now, he had failed to meet his famous client's only demand.

That night, Weintraub hardly slept. He got up early and went over to the convention center, thinking that he might be able to come up with something to salvage his career. He paced around the arena until an idea hit him. It wasn't possible to sell five thousand tickets to a concert in a few hours, especially if the concert were in a sauna. It wasn't even possible to *give away* five thousand tickets within a few hours. But he also realized that this was the wrong way to think about the problem: He didn't have to fill ten thousand seats; he just couldn't have any empty ones.

Weintraub remembered driving past a jail on the way to the convention center. He went there, found the sheriff in charge, and shoved a wad of money into the man's pocket. Weintraub told him that he wanted to remove five thousand seats from the convention center—

now—stow them, and then put them back before the evening performance. The sheriff was, of course, happy to help. He immediately put a detail of prisoners to work on it.

Weintraub had pulled it off. Both performances went well—Elvis hadn't noticed a thing, although he commented that there seemed to be more energy in a house for evening shows.

Jerry Weintraub was fortunate. His career-saving insight came just when he needed it, not the day after the matinee while looking for a new job. Most people, though, have insights at unpredictable, even inconvenient, times and places.

Andrew Cohen, a Detroit hip-hop DJ known as "Haircut," was driving to a party one night in 2005 when a new love song, both melody and lyrics in their entirety, suddenly occurred to him. Clearly this wasn't the best time to have an insight, because he couldn't write down the song while driving. Fortunately, another insight salvaged the situation: He sang the song into his cellphone and saved it as a message in his voice mail. "Just Ain't Gonna Work Out" became a breakout single in his new incarnation as the *Billboard*-charting artist Mayer Hawthorne.

Insight's unpredictability is fascinating. Grammy award–winner Taylor Swift shares, "When I write songs, it's never a conscious decision—it's an idea that floats down in front of me at four in the morning or in the middle of a conversation or on a tour bus or in the mall or in an airport bathroom. I never know when I'm gonna get an idea and I never know what it's gonna be."

Clearly, insights are mercurial. How can one reliably produce enough of them to study?

THE CHASE BEGINS

About a decade ago, we first collaborated to bring different neuroimaging techniques to bear—John specializes in EEG, Mark in fMRI—on the question of what happens in the brain during an aha moment. However, we faced a technical challenge: We couldn't follow people around 24/7 and wait for them to have epiphanies. To discern what happens in the brain during aha moments, we would have to bring people into the lab and reliably induce insights—*lots* of insights—rather than wait for a few of them to occur. Neuroimaging requires a large number of brain events. We needed specially designed problems and laboratory protocols to make this happen.

"PROBLEMATIC" THINKING

Consider the following three words: "pine," "crab," and "sauce." Can you think of a fourth word that will make a compound or familiar phrase with each of these words? Pause here and give it a try.

How might you solve this puzzle? The analytic approach is to consciously search through the possibilities and try out potential answers. For example, start with "pine." Imagine yourself thinking: *What goes with "pine"? Perhaps "tree"? "Pine tree" works. "Crab tree"? Hmmm . . . maybe. "Tree sauce"? No. Have to try something else. How about "cake"? "Crab cake" works. "Cake sauce" is a bit of a reach but might be acceptable. However, "pine cake" and "cake pine" definitely don't work. What else? How about "crabgrass"? That works. But "pine grass"? Not sure. Perhaps there is such a thing. But "sauce grass" and "grass sauce" are definitely out. What else goes with "sauce"? How about "applesauce"? That's good. "Pineapple" and "crab apple" also work. The answer is "apple"!*

This is analytical thought: a deliberate, methodical, conscious search through the possible combinations. But this isn't the only way to come up with the solution. Perhaps you're trying out possibilities and get stuck or even draw a blank. And then, "Aha! Apple" suddenly pops into your awareness. That's what would happen if you solved the problem by insight. The solution just occurs to you and doesn't seem to be a direct product of your ongoing stream of thought.

The "pine"/"crab"/"sauce" problem is an example of a type of puzzle called a "remote associates problem," developed by psychologist Sanford Mednick in the early 1960s for a test of creativity. In these little puzzles, the solution word is remotely associated with the words in the triplet. For example, when you hear the word "pine," the first associations that occur to you are probably words such as "tree" and "cone," rather than "apple." That's why the answer isn't immediately obvious.

Remote associates problems are short, and when people solve them, it usually takes them only a few seconds. This meant that we could get dozens of solutions from each participant. Another important benefit is that these problems can be solved either insightfully or analytically. This allowed us to compare brain activity for these two ways of solving without changing the type of problem. As it turned out, almost all of our participants solved some of the puzzles analytically and some with insight, which makes sense because, in real life, almost everyone uses both modes of thinking from time to time.

To compare insight and analytic solving, we needed to know which problems a person solved with each of these strategies. We found this out simply by asking them. After a participant solved each puzzle, he reported whether the solution had popped into his awareness (insight) or whether it had occurred to him by consciously evaluating potential solutions (analysis). Knowing how each person solved each puzzle allowed us to compare brain activity for insight and analytical solving while holding constant all the other types of processing done for the puzzles—for example, reading the words and pressing a button.

But how could we know that our research participants were solving these problems the way they said that they were? If participants couldn't be trusted to correctly report how they came up with each solution, then the EEG and fMRI results would be useless. We have several reasons to be confident about people's ability to correctly report how they solve such problems. One of these lines of evidence has implications that go beyond the laboratory.

FACING DOWN THE CLOCK

Almost all of our participants solved some problems analytically and some insightfully. But they differed in the mix of these two strategies. Some solved more with insight, others more by analysis. We compared these two groups of people and found that they tended to make different kinds of mistakes. Insightfuls made more "errors of omission." When waiting for an insight that hadn't yet arrived, they had nothing to offer in its place. So when the insight didn't arrive in time, they let the clock run out without having made a guess. In contrast, Analysts made more "errors of commission." They rarely timed out, but instead guessed—sometimes correctly—by offering the potential solution they had been consciously thinking about when their time was almost up.

The fact that the Insightfuls tended to time out and make errors of omission while the Analysts often guessed and made more errors of commission shows that people's judgments about their problem-solving strategies weren't random. Their judgments accurately reflected the strategy that they used. This helped to validate our experimental technique, but it is more than a technical point about our study. It reveals an important truth about problem solving outside the laboratory: Analytic problem solving is amenable to deadlines; creative insight is not.

If you're consciously and methodically working on a problem and you're put on the spot to respond *now,* you can guess by offering up your work in progress. It's probably better than nothing. It may even be good. On the other hand, insights can't be forced. The unconscious thinking that culminates in a conscious insight doesn't care about your deadline. If an answer is demanded of you before you've had your epiphany, then you won't have an answer to give. The takeaway point is that you should neither expect nor demand outside-the-box thinking—from yourself or anyone else—on a strict timetable. Providing some structure can help to keep people on task, but creativity needs elbow room.

At any rate, we had worked out the details of the problems our participants would try to solve and procedures for isolating their insights. We were now free to apply our neuroimaging tools to isolate aha moments in the brain.

WHEN AND WHERE

We used two complementary techniques for measuring brain activity. EEG gives precise measurements of the timing of electrical brain activity—the "when"—with less-precise information about what parts of the brain are generating these signals. Functional magnetic resonance imaging, or fMRI, shows precise locations of brain activity—the "where"—but with less precision about its timing. Both the when and the where proved to be essential ingredients.

In an fMRI experiment, a person reclines on a table that slides into a scanner containing a large, doughnut-shaped magnet (see figure 5.1). Different regions of the brain perform different functions, and a region draws more blood when it's doing more work. The scanner measures the amount of blood flowing to each area because of the

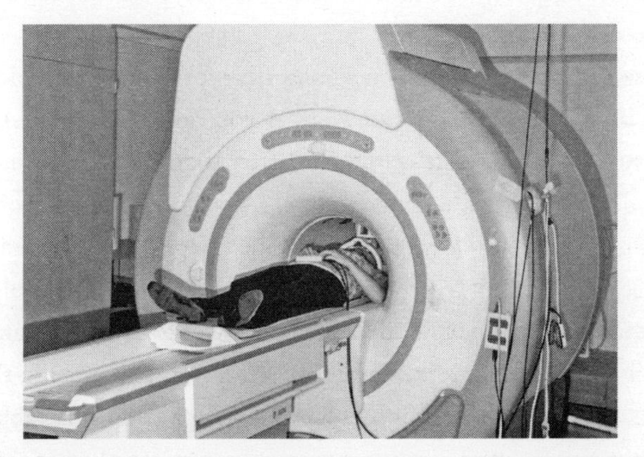

FIGURE 5.1: A functional magnetic resonance imaging (fMRI) scanner. *Picasa (picasaweb.google.com/lh/photo/4hLOTdOOnq6kL73rJsiMAA)*

magnetic properties of the iron-rich hemoglobin in the blood. The result is a three-dimensional map of the brain showing which areas are most active.

Functional magnetic resonance imaging can tell us only roughly when a brain event occurs because it takes several seconds for blood to rush to a cranked-up brain area. In some cases, fMRI will even show blood flowing to an area seconds after the neural event ended. What fMRI measures is therefore a shadow of the neural activity underlying thought. That's the price we pay for its exquisitely precise maps of brain areas at work.

In our experiment, we measured these two aspects of neural activity—blood flow, using fMRI, and electrical current, using EEG—when people solved remote associates problems by insight and by analysis. By comparing these two ways of solving puzzles, we hoped to discover which brain areas were more active in each. The timing and location of brain activity were both critically important to our goal. We wanted to localize brain events occurring right at the aha moment and not confuse them with other neural activity leading up to or following the insight.

We found that at the moment a solution pops into someone's awareness as an insight, a sudden burst of high-frequency EEG activity known as "gamma waves" can be picked up by electrodes just above the right ear. (Gamma waves represent cognitive processing in the brain, such as paying attention to something or linking together different pieces of information.) We were amazed at the abruptness of this burst of activity—just what one would expect from a sudden insight. Functional magnetic resonance imaging showed a corresponding increase in blood flow under these electrodes in a part of the brain's right temporal lobe called the "anterior superior temporal gyrus" (see figure 5.2), an area that is involved in making connections between distantly related ideas, as in jokes and metaphors. This activity was absent for analytic solutions.

Additional work led us to update our findings. A few years later,

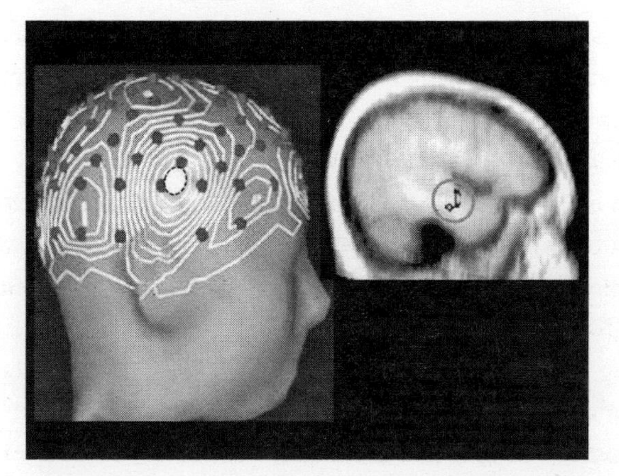

FIGURE 5.2: On the left, an EEG map of the right side of the head showing the location of the insight gamma-wave burst (the white oval just above the right ear). The gray dots show the locations of the electrodes. On the right, an fMRI image of the corresponding increased blood flow to the right temporal lobe (shown inside the gray circle). *PLoS Biology, 2, 500-510. (www.plosbiology.org/article/info%3Adoi%2F10.1371%2Fjournal.pbio.0020097)*

we replicated the fMRI study using more powerful procedures and found that the right temporal lobe isn't the only part of the brain associated with insight. A whole network of brain areas is involved. In fact, all types of thinking require complex networks of brain areas. But the burst of activity in the right temporal lobe was still the most prominent and reliable effect that we found, and it was the only effect that our original EEG study showed to occur right at the moment of insight.

So we had found a neural signature of the aha moment: a burst of activity in the brain's right hemisphere. Almost literally, this is the spark of insight. We were excited by this result, but we weren't entirely surprised. Earlier research had revealed why the brain's right hemisphere is usually the point of origin for creative ideas.

6

THE BEST OF BOTH WORLDS

The notes I handle no better than many pianists. But the pauses between the notes—ah, that is where the art resides.

—Artur Schnabel, pianist

D.B. was a lucky man. His neurologist told him so.

A trial attorney and avid runner, D.B. was just forty-nine years old and in excellent health when he suffered a stroke—an artery had burst and bled into his brain, depriving a chunk of brain tissue of oxygen. This stroke occurred at the junction of the temporal and parietal lobes, just behind his ear, damaging the "white matter" through which brain areas communicate with one another. This cut off some of the "gray matter" of the cerebral cortex from communicating with the rest of his brain.

What made D.B. "lucky" was that the stroke had damaged his right hemisphere rather than his left. Had the stroke occurred in the mirror-image left-hemisphere region, he would have experienced Wernicke's aphasia, a profound deficit of language comprehension.

In the worst cases, people with Wernicke's aphasia may be completely unable to understand written or spoken language.

Brain damage can cause any of several types of aphasia, each with its own pattern of symptoms. Despite this variety, the vast majority of aphasics have one important thing in common—the damage is almost always to their left hemisphere. But D.B.'s left hemisphere remained intact, leading his neurologist to tell him not to expect any language difficulties. D.B. had dodged a bullet.

Mark met D.B. two years after his stroke and, on the surface, D.B.'s language ability did seem fine. He spoke fluently. He answered questions well. He spoke clearly, at a normal rate, and without any unusual intonation or accent. D.B. had taken some standard tests of aphasia to examine his ability to perceive sounds, understand words and sentences, and answer questions with well-formed responses. He was able to do all of these things. According to the tests, he had no language deficits.

Nevertheless, D.B. didn't feel lucky. He may have been better off than if he'd had a left-hemisphere stroke, but he felt that his language ability was far from normal. He said that he "couldn't keep up" with conversations or stories the way he used to. He felt impaired enough that he had stopped litigating trials—he thought that it would have been a disservice to his clients to continue to represent them in court.

The paradox of this stroke patient who didn't seem to have aphasia yet complained of language difficulties fascinated Mark and spurred him to investigate the language abilities of the right hemisphere. At the time, he didn't realize where this would eventually lead. Following up on prior research by psychologists Hiram Brownell, Howard Gardner, and others, he compared D.B. and other right-hemisphere-damaged patients to a group of healthy control participants. D.B. and the other patients were able to understand the straightforward meanings of words and the literal meanings of sentences. Even so, they complained about vague difficulties with language. They failed to grasp the gist of stories or were unable to follow multiple-character or multiple-plot stories, movies, or television shows. Many didn't get jokes. Sarcasm and irony left them blank with incomprehension. They could sometimes muddle along without these abilities, but whenever things became subtle or implicit, they were lost.

To probe these symptoms, Mark played audio recordings of simple, brief stories for the patients and the healthy participants. The stories contained little episodes such as the following: Saturday, Joan went to the park by the lake. She was walking barefoot in the shallow water, not knowing that there was glass nearby. Suddenly, she grabbed her foot in pain and called for help, and the lifeguard came running.

After listening to each story, they answered true/false questions, half of which concerned simple facts from the stories. Here are a couple of examples of straightforward questions about the Joan story:

True or false?

Joan went to the park on Saturday.

The park was near a river.

D.B. could obviously hear the words, understand their basic meanings, and remember the stories well. In fact, he answered 100 percent of these kinds of factual questions correctly—a feat not achieved by any of the 120 healthy college students who also took this test. Still, he complained: "Even though I understand the words, I no longer understand the subtleties, the complex mosaic of meaning that is language." Such eloquence would seem to suggest excellent language ability. Yet, he was very perceptive. He was indeed missing something.

The other half of the true/false questions focused on events that were only implied by the stories. These required the kinds of simple inferences that healthy people make quickly and effortlessly, even automatically. For example:

True or false?
Joan almost drowned.
Joan cut her foot.

After hearing that Joan had walked near glass and grabbed her foot in pain, most people assume that she had cut her foot. In fact, if you ask people whether they actually heard "Joan cut her foot," many will mistakenly say "yes." As we showed in a previous chapter, our brains use their mental boxes to automatically fill in the gaps to make sense of things.

D.B., who had answered the factual questions better than all of the control subjects, nevertheless flubbed simple inferential questions that the average college student could answer with ease. For example, he incorrectly answered "true" when asked whether Joan had almost drowned and "false" when asked whether Joan had cut her foot. And when asked to recall the stories, he remembered the stated facts without these implied events. He was oblivious to the holes in the narratives.

To clarify this deficit further, Mark asked the participants to listen to little stories while they kept their eyes fixed on a computer screen. Whenever a word appeared on the screen, their task was simply to say it as quickly as possible. After hearing "Joan walked near glass, then grabbed her foot in pain," healthy people who saw the word "cut" could say this word quickly—more quickly than others who saw "cut" after hearing a sentence that had nothing to do with glass, feet, or pain. This showed that at some level the healthy participants inferred that Joan had cut her foot on the glass. But D.B. and his fellow patients always responded more slowly to "cut"—it didn't matter whether the preceding story had been about Joan walking near glass or something unrelated. They didn't make the connection between "cut" and Joan's foot.

You might say that D.B. and the other patients couldn't read between the lines. In fact, they couldn't understand the phrase "read between the lines." This strange inability turned out to be a major clue to understanding the right hemisphere's role in insight.

WHAT'S THE DIFFERENCE?

It became apparent that the classical neurology view that only the left hemisphere processes language was limited. The left hemisphere clearly doesn't have a monopoly on language. But then what is the role of the right hemisphere? How does it make inferences, derive the gist of a story, and understand jokes and metaphors—all the things D.B. couldn't do?

The work of cognitive neuroscientist Christine Chiarello provided another important piece of the puzzle. She uses the "visual half-field" technique to present information to one hemisphere at a time. Don't worry—this isn't invasive. When you look straight ahead, information about objects to the left of your eye gaze goes directly to your right hemisphere and information about objects to the right of your gaze goes to your left hemisphere. (The nerves cross over.) Of course, once the information goes to one hemisphere, it's quickly communicated to its partner through connecting nerve fibers. But the hemisphere that gets the information first has a head start (pun intended) that allows it to dominate.

Chiarello's studies showed that both hemispheres store the meanings of words. To those steeped in the classical view, that was surprising enough. Her key discovery, though, was that the hemispheres store different types of associations between these words.

The left hemisphere is sharp, focused, and discriminating. When a word is presented to the left hemisphere, the meaning of that word is activated along with the meanings of a few closely related words.

For example, when the word "table" is presented to the left hemisphere, this might strongly energize the concepts "chair" and "kitchen," the usual suspects, so to speak. In contrast, the right hemisphere is broad, fuzzy, and promiscuously inclusive. When "table" is presented to the right hemisphere, a larger number of remotely related words are weakly invoked. For example, "table" might activate distant associations such as "water" (for underground water table), "payment" (for paying under the table), "number" (for a table of numbers), and so forth.

So it became clear why a person relying on only one hemisphere would have trouble with particular aspects of language. Consider "Joan walked barefoot in the shallow water, not knowing there was glass nearby." With only a right hemisphere, a person would have trouble figuring out whether the nearby "glass" referred to a submerged shard or a drinking glass. In contrast, the left hemisphere would know what kind of glass was involved but couldn't fill in the gap between "barefoot" and "glass" to infer that Joan had cut her foot.

Neuroimaging studies also show that people frequently consult their right hemispheres while processing language. Hot spots in left-hemisphere language areas are often accompanied by weaker, mirror-image shadows in the right hemisphere. This right-hemisphere activity can be scaled up when a person needs to comprehend the remote associations in jokes and metaphors, when reading stories that require filling in gaps, or when reading a story that doesn't have a title to clue in the reader about the theme of the plot. And these right-hemisphere remote associations aren't just an add-on. They provide a backup interpretation that kicks in when the left hemisphere just doesn't get it. In particular, jokes are designed to cause this kind of initial misinterpretation followed by a switch to the remote alternate meaning at the punch line. (As Henny Youngman used to say, "My grandmother is over eighty and still doesn't need glasses. Drinks right out of the bottle.")

STEPPING UP

Are you now convinced that the two hemispheres think about the world in radically opposing ways? You shouldn't be. The fact that the two hemispheres have different strong suits doesn't mean that they are fundamentally alien to each other. To the contrary, a remarkable phenomenon shows that the two hemispheres are cut from the same cloth.

Young children with Rasmussen's syndrome have debilitating seizures limited to one hemisphere. When medications fail to control the seizures, the patients may need a "hemispherectomy"—the entire affected hemisphere is surgically removed. This treatment is radical, but it's necessary when the seizures are so intense that the children are otherwise disabled. The astonishing thing is that many of the Rasmussen children whose left hemispheres have been removed eventually end up with normal or near-normal language ability. In time, the right hemisphere can pick up most of the slack.

The brain's ability to remodel itself—its "plasticity"—is greater in children than in adults. Nevertheless, even adults with severely damaged left hemispheres can partially, sometimes even fully, recover their language ability in the months and years following their injuries. In some of these cases, their recovery is at least partly due to contributions from the right hemisphere. We know this from patients who became aphasic following a left-hemisphere stroke. They slowly recovered and then suffered more severe aphasia following a second stroke that damaged the right hemisphere, proving that their right hemispheres had taken over some of the left hemisphere's language abilities. Neuroimaging studies have also shown that recovery from a left-hemisphere stroke can lead to increased activity in the right hemisphere.

This is all compelling evidence that the left hemisphere doesn't have exclusive rights to language. The right hemisphere not only complements the left's language ability; when necessary, it can also replace it. This couldn't happen if the two hemispheres were fundamentally alien to each other.

Nevertheless, there are dissimilarities, and these may account for the different views of the world from the left and the right sides of the brain.

IT TAKES ALL KINDS

Though every neuron in your brain receives and combines information from other neurons, there is considerable variety: Some neurons accept inputs from many neighbors, some from fewer. Those that receive messages from a small number of neighbors tend to be rather specialized. For example, some of your visual-system neurons fire only when you see a particular feature, such as a line or an edge. Some are yet pickier—they will respond only to a specific combination of features, such as a line at a particular angle and in a particular location. This kind of pickiness enables you to distinguish one object from another. Without it, you couldn't tell a baseball from a tennis ball. On the other hand, the nonchoosey neurons help you see what a baseball and a tennis ball have in common: They are both balls.

The idea is that neurons that process words and ideas aren't all that different from those that accomplish vision. Some language neurons receive inputs from a small number of neighbors and may make fine distinctions between concepts. Others accept a wide range of inputs and may form remote associations between ideas. Importantly, choosey and nonchoosey neurons tend to be segregated.

A brain looks pretty symmetrical to the naked eye, but if you look at samples of the two hemispheres under a microscope, you'll see that there are subtle differences. Left-hemisphere language areas have more of the picky neurons with the exclusive social circles; right-hemisphere language areas have more of the neurons with expansive social lives. We don't yet know for sure whether this causes the two hemispheres to process concepts differently. It's a huge leap from observations about neuronal circuitry to mental associations. Nevertheless, this difference in circuitry must somehow influence the way the hemispheres process information, and the cognitive styles of the two hemispheres must have some basis in the brain's circuitry.

If borne out by further research, this understanding of neuronal circuitry has an important implication: The hemispheres really aren't all that different from each other. Just as you can tune two radios to slightly different frequencies to pick up wildly different radio stations, a slight difference in neuronal tuning can empower the hemispheres to think about the world in diverse ways: focused and exclusive or broad and fuzzy.

RIGHT ON!

Since remote associations are the stuff of insights, and the right hemisphere is the home of remote associations, it made sense to go one step further to predict that the right hemisphere is also the origin of insight. Edward Bowden and Mark tested this idea in several visual half-field studies before the neuroimaging work described in the last chapter. They hypothesized that an aha moment occurs when an idea that's already slightly activated in the right hemisphere—but is still unconscious—suddenly emerges into awareness as an insight.

Each participant viewed three-word remote associates problems such as those introduced in the previous chapter (for example, "fish"/"mine"/"rush"; solution: "gold" for "goldfish"/"gold mine"/ "gold rush"). After each problem, a fourth word was flashed to the left or right hemispheres of the participants. Their task was simply to say that fourth word out loud as quickly as possible. Sometimes the fourth word was the solution to the problem ("gold"), and sometimes it was unrelated (for example, "rock"). Ed and Mark reasoned that if a problem unconsciously activates its solution before a participant has consciously realized the solution, then he or she should say the fourth word more quickly when it's the solution than when it's unrelated to the problem. This turned out to be true—but only when the word was flashed to the right hemisphere.

This result showed that the right hemisphere plays an important role in solving these problems. All well and good. However, by itself, it didn't prove that the right hemisphere plays a role in solving them by insight. Another finding provided this proof: People said the solution words faster when presented to the right hemisphere—but only when these solution words evoked an aha experience!

When the presentation of a problem activates its solution through the remote associations of the right hemisphere, the solution isn't energized enough to break through to awareness. However, when the solution word or a relevant hint is projected directly to the right hemisphere, this gives the idea just enough of a bounce to enable it to bubble up as an aha (see figure 6.1).

This finding, together with all the research on right-hemisphere language processing, laid the foundation for our first neuroimaging

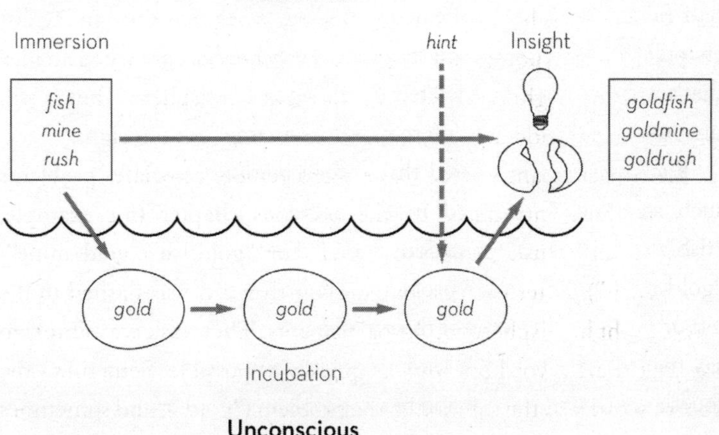

FIGURE 6.1: The presentation of a problem can unconsciously activate its solution in the brain's right hemisphere. But sometimes it takes an external hint to energize this solution to pop into awareness as an insight.

study of insight. That's why we weren't surprised when we found that insights are accompanied by a burst of neural activity in the right hemisphere.

However, before moving on to discuss factors that promote insight, a final thought puts this research into its proper perspective. The right hemisphere's facility for remote associations and alternate interpretations makes it critical for insight, but real-life accomplishment often requires both remote and close associations, both insightful and analytic thought. In the world outside of the lab, insights may have to be evaluated, verified, refined, and applied, and this requires contributions from the more analytic left hemisphere. Just as your ability to use language requires two intact hemispheres, so does effective, practical, creative performance. But the right hemisphere provides the spark that ignites the creative fire.

7

TUNING OUT AND GEARING UP

I close my eyes in order to see.

—Paul Gauguin, artist

During a video tour of Pixar Animation Studios' creativity-enhancing facilities, John Lasseter and Pete Docter, who have written, directed, and produced blockbuster Pixar movies, sat in curious chairs with domes that resemble the kind of hair dryers found in beauty salons. After they pulled the domes over their heads to isolate themselves from their surroundings, Lasseter said, "This is where we get our creativity." Docter, abruptly looking insightful, pointed up and added, "Oh, I just had an idea there." Lasseter and Docter were, of course, joking—at least about the use of the chairs at that moment. But their whimsical presentation conveys an important truth about creativity. The glare of the external world can block insights from emerging. Sometimes you have to isolate yourself and focus your attention inwardly to allow a new idea to surface. The brain can achieve this state of inward focus in several ways.

THE IDLING BRAIN

Our first neuroimaging study, which we described in Chapter 5, produced a finding that took us a while to understand. Recall that at the moment of insight there is a burst of EEG gamma waves in the right hemisphere. About a second before that, there is a burst of EEG alpha-wave activity measured on the right side of the back of the head (see figure 7.1).

When neurons fire at the slower alpha frequency, they aren't actively processing information. A useful analogy is that of idling your car by shifting the transmission into park. The car is working, but it isn't going anywhere. Alpha is a neuron's park.

So, just before an insight, a region in the back of the brain downshifts into alpha. Then the right temporal lobe fires up into the

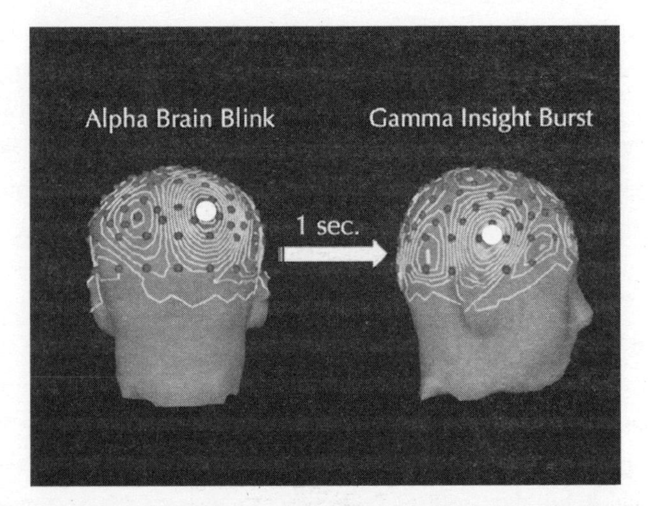

FIGURE 7.1: An alpha "brain blink" (white circle at the back of the head) precedes the gamma insight burst (white circle on the right side of the head). Each dark dot shows the location of an EEG electrode. *John Kounios and Mark Beeman*

gamma rhythm to proclaim the solution word that links together the words of the remote associates problem. That's the aha moment.

This might seem paradoxical at first: How could briefly idling a part of the brain help set the stage for an insight? Shouldn't creativity require doing something extra rather than something less? As we'll see, sometimes you have to stop doing one thing before you can start doing something else.

BLINKING THE MIND'S EYE

In the 1920s, German neurologist Hans Berger discovered that when a person sits quietly with his eyes closed, prominent alpha waves can be recorded at the back of his head right over the brain's visual processing areas. When he opens his eyes, this alpha suddenly disappears. Berger understood that these alpha waves are a sign of reduced visual perception—more alpha means less intake of visual information. When the visual part of the brain doesn't get much to work with, it idles at the slow alpha frequency. But when you open your eyes, the visual cortex is reengaged and begins humming at higher frequencies to process the incoming rush of information.

We didn't expect a burst of alpha in our experiment, and we didn't immediately know what to make of it. (Our participants weren't closing their eyes to create the alpha.) Appropriately, an understanding of this finding came to John as a sudden realization.

Consider what happens when someone asks you a difficult question. You don't know the answer immediately and have to give it some thought. To make this easier, you might look away from your questioner's distracting face or even close your eyes to focus more on your thoughts. However, the brain has another way to avoid such distractions: by temporarily taking in less visual information. Humans are inherently visual creatures. A huge portion of the brain is specialized for visual cognition; vision often dominates and overshadows other types of thought. That's not to say that vision must be

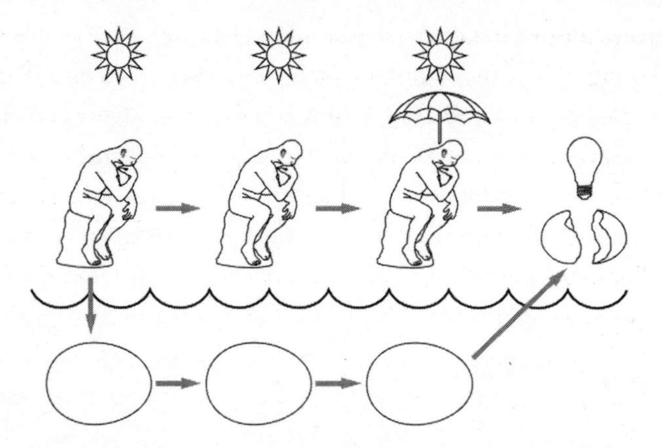

FIGURE 7.2: Removing the glare of the external world can allow insights to emerge.

turned completely off. Andrew Stanton's *WALL-E* insight shows otherwise. But sometimes it's necessary to dial down the dimmer so that you won't become blinded to subtler thoughts.

This happened to John. He was poring over a printout of the data from the experiment and saw the burst of alpha waves. At first, it made no sense. He thought about it for a while longer but still had no idea what it meant. Then he looked away from the papers that were spread out on the table. His gaze drifted up to the blank ceiling of the laboratory. He thought: *Alpha burst precedes insight. . . . Hans Berger . . . alpha as brain idling . . . regulation of visual inputs . . . Insights occur when subconsciously activated ideas pop into awareness. . . .* Then, *aha!* It all came together and made sense: The alpha burst is a "brain blink" that briefly cuts off visual inputs to the brain to reduce distraction. This allows one's attention to find the new idea and jolt it into consciousness. Quite literally, John's reducing his visual inputs by looking up at the blank ceiling enabled an insight into the very process that he was trying to understand. This was an insight into insight.

OK, there is an alternative explanation for this alpha burst that we'll mention and then dismiss. Instead of catapulting an unconscious idea into awareness, could the brain blink somehow enable the mind to figure out the solution from scratch right then and there? Could closing your mind's eye cause a new idea to pop into existence rather than enable a preexisting idea to pop into awareness? It's an appealing idea that does seem to happen in some situations, but it's rather implausible here. The alpha blink in our experiment occurred only about a second before each insight. That's just not enough time for someone to figure out the answer to one of these rather challenging puzzles from scratch. In this case, the solution must already be ready and waiting—unconsciously—when the blink enables you to find it.

This alpha brain blink seems to be a quick, automatic mechanism that you probably can't control for your benefit. But that doesn't mean that you can't achieve a similar effect the old-fashioned way.

A brain blink helps by reducing visual distractions so that it's easier to find and lock on to an insight. Fortunately, longer periods of sensory restriction can accomplish the same thing. It's become a cliché, but many people have reported that they have had some of their best insights in the shower—perhaps the modern equivalent of Archimedes's bath. The white noise of the running water is hard to focus on and blocks out other kinds of sounds. The warm water makes it difficult for you to feel the boundary between the interior and exterior of your body, so your sense of touch recedes from awareness. The visual inputs are unchanging and blurry. Perhaps your eyes are even closed to keep the soap out. Taking a shower is an excellent way to cut off the environment, focus your thoughts inwardly, and have an insight. Aaron Sorkin, the creator of the television series *The West Wing* and writer of the movie *The Social Network,* said, "Writer's block is like my default position. When I'm able to write something, that's when something weird is going on." His technique for removing writer's block includes taking six or more showers per day. However, bathing isn't the only way to disengage from the outer world. The acclaimed writer Jonathan Franzen used more extreme measures while working on his novel *The Corrections*. To coax his imagination, he would often type in the dark while wearing earplugs, earmuffs, and a blindfold. Whatever works.

INSIGHT AND OUTSIGHT

Louis Pasteur, the great pioneer of biomedical research, once said, "Chance favors only the prepared mind." This statement is a bit ambiguous even in its original context. We interpret "prepared mind" to mean a specific brain state that inclines one to solve problems by insight. Clearly, the existence of such a brain state would be an important discovery, not only because it would yield evidence about the origins of insight, but also because it would suggest ways to spark aha moments. In a neuroimaging study we did with cognitive neuroscientists Jennifer Frymiare Stevenson and Jessica Fleck, we looked for, and found, such a state.

We proposed that a person's brain state would incline her to tackle a problem either insightfully or analytically even before she knows what the problem is. As in our previous neuroimaging study, participants attempted a series of remote associates problems. This time, instead of measuring their brain activity while they solved each problem, we measured it during the two seconds just before each puzzle was displayed. The experiment showed a striking pattern of brain activity underlying a prepared mind.

Just before a person sees a problem that she will eventually solve by insight, there is greater activity in the temporal lobes of both hemispheres (see figure 7.3). By itself, this wasn't a big surprise. The temporal lobes, which are near the ears, are involved in processing words and concepts. When a person prepares to solve a verbal puzzle, it makes sense that her brain would get ready by activating its knowledge of words and their meanings. It also makes sense that the temporal lobes of *both* hemispheres would be primed—this would energize both close and remote associations. But it wasn't immediately obvious why the temporal lobes would be more active before tackling a problem insightfully than analytically. After all, whichever mind-set you use, you still have to think about the meanings of words.

The difference is that for a person in an insightful frame of mind, everything is up for consideration. Nothing is off the table. Any idea—*every* idea, no matter the source—is considered a potential solution. That's why both the left and right temporal lobes light up like a Christmas tree when a person adopts an insight mind-set. It's the neural manifestation of openness to the full range of possibilities. But how does this state of openness come about?

Think back to elementary school. Remember how your teacher would pose a question and then pause to observe which students would raise their hands to offer an answer? Typically, one assertive student would shout out, "Ooh! Ooh! I know the answer!" and a few of the less-aggressive ones in the back of the room might meekly raise their hands halfway up. The teacher had to decide whether to call on the lone show-off again—he always raises his hand ostentatiously—or whether to call on one of the shy students in the back to answer the question. The key to this analogy is that teachers often make the decision to call on a shy student even before they pose a question to the class. And when the shy students realize that the teacher is encouraging them to speak up, then more of them become willing to raise their hands to offer an answer.

When you encounter a problem, it can evoke several potential solution ideas. Often, there are one or two strong, obvious possibilities, such as running away from the Mann Gulch fire that confronted Wag Dodge. But other ideas, such as using fire as a tool for creating a protective buffer zone, aren't obvious because they are weakly associated with the elements of the problem. Fighting fire with fire isn't the first thing that jumps to mind in such situations. These long-shot ideas are the shy students, sitting at their temporal lobe desks.

To continue the analogy, the teacher is a brain area called the "anterior cingulate" and located in the middle of the front of the head

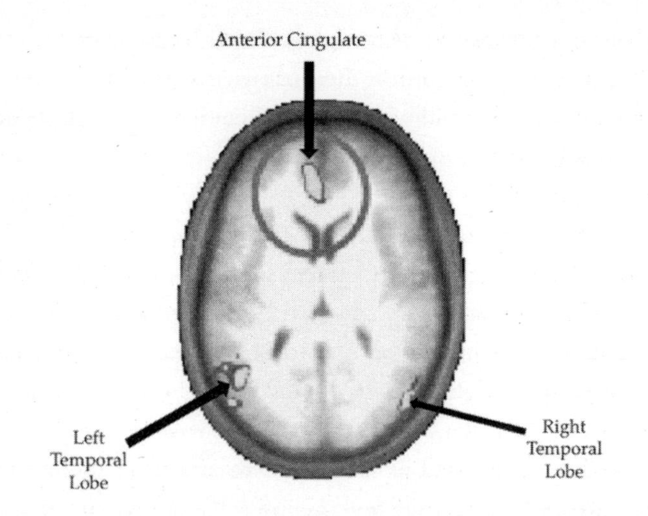

FIGURE 7.3: An fMRI image showing areas of the brain that were more active when preparing to solve a problem by insight. This is a view of the top of the head, with the back of the head at the bottom, the front of the head at the top, etc. The circled area shows insight-related activity in the anterior cingulate. Other areas associated with preparation for insight include the left and right temporal lobes.

(see figure 7.3). We proposed that this area monitors the rest of the brain for the presence of competing ideas, especially the quiet, nonobvious ones. If the anterior cingulate detects the presence of shy alternative possibilities, then they won't get squashed or overshadowed by the assertive ones and they can remain in play as contenders. Thus, the anterior cingulate is a key contributor to the insight mind-set—when this brain region is active, you are open to a broad range of potential solutions, even shy ones that don't seem particularly promising at first.

The flip side of this is the analytic mind-set. When your anterior cingulate is relatively inactive, you experience a form of mental tunnel vision: One or two obvious ideas upstage the shy ones.

While in the analytic mind-set, our participants also showed widespread activity in visual parts of the brain. This suggests that they were looking intently at the screen on which the next problem was about to be displayed. We might call this "outsight"—the outward focus of attention on the source of the problem rather than on one's own thoughts and ideas. It's the domination of thought by the straightforward appearance of things.

Most of the problems we encounter in our day-to-day lives can be taken at face value. You don't always need to look for obscure solutions. Usually, you can afford to ignore the weak, conflicting voices that are whispering alternate possibilities to you, because the loud voice shouting something right in your face is most often right. This isn't creative or subtle, but it's usually effective. This is the way of the Analyst.

So we have these two basic mind-sets for solving problems—insight and analytic. Shifting your attention more inwardly or outwardly can help you to engage one or the other. But which should you choose? Analysis is easier to trigger and is available on demand. Most everyday problems don't need creative solutions anyway. It therefore makes sense to start with analysis. But in situations calling for outside-the-box ideas—that is, when you've exhausted the possibilities—you can use sensory deprivation to help you focus inwardly. The missing ingredient needed to realize a great idea could be as simple as a little solitude and inner attentiveness to minimize the distractions that might be blocking an idea from bubbling up. After all, you can't see the stars when the sun is out.

8

THE INCUBATOR

Even a soul submerged in sleep is hard at work and helps make something of the world.

—Heraclitus, *Fragments*

Many people have reported waking up with an idea or the solution to a problem that they incubated during sleep. Some people even keep a notepad next to their bed to document their fleeting thoughts.

Some people keep a piano.

One morning in 1964, Paul McCartney woke with a melody playing in his head—he had heard it in his dream. "The tune itself came just complete, came just out of a dream," he explained. He immediately got up and went to the piano next to his bed and played through it. He added words later to create the song "Yesterday," which several polls, audiences, and music experts have voted the greatest pop song of the twentieth century.

"If you're really lucky they just arrive and you kinda just write 'em down," said Sir Paul.

And it is very important to write them down—*legibly*.

HOTBED OF INCUBATION

One night in 1921, the German pharmacologist Otto Loewi was awakened by a dream about a simple experiment that he was certain would resolve a pressing scientific question of the day. He scribbled down the idea and went back to sleep. The next morning, he examined his notes but could neither read his writing nor remember the specifics of his dream. Fortunately, he was awakened by the same dream the next night. This time, rather than risk being left with undecipherable scrawl again, he went directly to his laboratory to perform the experiment.

Early twentieth-century neuroscientists had a rudimentary grasp of the brain's basic structure. But their understanding was limited by a mystery. They knew that individual neurons send signals to one another. They just didn't know how. The problem is that neurons don't touch. A tiny gap called the "synapse" separates the business end of a transmitting neuron from a target neuron. The question was how a signal passes across the synapse.

There were two possibilities. One was that an electrical signal jumps across this gap the way a bolt of lightning passes through air. The other idea was that the transmitting neuron secretes some kind of chemical that another neuron picks up from the synapse. Though these two scenarios are quite different, early twentieth-century technology didn't seem advanced enough to allow researchers to prove one or the other. Loewi's dream yielded an astonishing, but simple, solution.

Loewi prepared two live, beating frog hearts, each immersed in a liquid in its own dish. He electrically stimulated the vagus nerve attached to the first heart, which caused that heart to beat more slowly. Then he extracted some of the liquid from around the connection between the nerve and the heart and applied it to the heart in the other dish. The other heart also started beating more slowly. This showed that the neurons in the vagus nerve released a chemical into the gap with the heart, slowing the beating of the second heart. Thus,

Loewi's sleep insight enabled him to demonstrate that the signal transmitted across a synapse is chemical, not electrical, in nature. For his dream-inspired research, Loewi shared the 1936 Nobel Prize in Physiology or Medicine. His discovery has led to great strides in understanding illnesses such as schizophrenia and Alzheimer's disease.

Instead of arriving during sleep, some insights arrive soon after a person awakens. The mathematician and philosopher René Descartes had the habit, and luxury, of staying in bed for some time after waking in the morning. One such morning, he looked up from his bed and saw a fly buzzing around. He asked himself how he could mathematically describe the position of the fly in space at any given moment. Then he had an insight: The floor and walls of his bedroom could be thought of as a three-dimensional system of axes (X, Y, and Z). At any given moment, the fly was a specific distance from the floor and the two walls. These three distance measurements could precisely specify the spatial location of the fly. Thus was born the coordinate system of analytic geometry.

Of course, Descartes's insight may have occurred simply because his mind was fresh in the morning. But there is another possibility. Researchers have discovered sleep's neural architecture: a series of stages, each with its own pattern of brain activity. Though distinct, each sleep stage has some "momentum." When you are awakened, your brain doesn't immediately snap into a fully awake state (a fact of which coffee drinkers don't need to be reminded). Aspects of sleep stages can persist for a while. Descartes's insight may have been the product of a conscious mind still under the partial influence of the last phase of the previous night's sleep. It's likely that he had been pondering his geometrical question, off and on, for weeks, months, or even years before he had his insight. If so, he may have been incubating this problem just before waking. This sleep incubation could have persisted as he lay in bed staring at the ceiling until the fly triggered his insight. Many people report experiencing insights shortly after awakening, as if from thoughts carried over from sleep.

SNOOZE, LOSE, AND REMEMBER

One way that sleep can incubate ideas and promote insight is by removing blocks that prevent a weakly activated idea from becoming conscious. Cognitive psychologists call this liberation of thought "fixation forgetting."

Early in his career, acclaimed journalist and author Pete Hamill worked as a newspaper reporter and columnist, though he wanted to write fiction as well. But whenever he would sit down to work on a novel, thoughts of the real-life events he wrote about as a journalist would emerge and stifle his imagination. He even had trouble escaping from his newspaper style of writing. Fortunately, he found a solution: "I would lie down, [and] think about . . . the fictional story I wanted to write. Passed out, the subconscious would help me to make the choices of what I was going to write, so I had energy when I woke up and I forgot about the newspaper work that I had done that day. . . . The creative uses of the nap are very much underrated," he mused.

Though examples of sleep-related creativity by McCartney, Loewi, Hamill, and many others are compelling, there is surprisingly little direct scientific evidence for problem incubation during sleep. Some tantalizing studies purport to have directly demonstrated sleep-related incubation, though we find them limited or ambiguous. (We discuss this in the notes to this chapter.) However, another line of research does provide solid evidence. It shows that sleep brings out the remote associations that are embedded in memories.

New memories are initially fragile. Over time, they become strengthened and stabilized by a process called "memory consolidation." Cement is a good analogy: Wet and soft when first poured, it is strong and durable a few hours later once it has hardened. However, unlike cement, memory consolidation can take place over decades. This is why a head injury is more likely to interfere with recent memories than old ones—the older memories have had years to "harden."

The cement analogy does have limits, though. Cement dries and hardens but otherwise doesn't change much. In contrast, consolidating a memory not only stabilizes it but can also transform it by bringing out associations that are only implied in the original version of the memory. Much of this takes place during sleep. For example, let's say that you learned the following facts:

Tom is taller than Bill.
Bill is taller than Jack.
Jack is taller than Phil.
Phil is taller than Steve.

Note that in addition to Tom being taller than Bill, he's also taller than Jack, Phil, and Steve, even though this wasn't explicitly stated. Neurologist Jeffrey Ellenbogen and colleagues found that people were better at remembering such implied facts after a nap, compared with people tested after a sleepless break. Sleep seems to highlight the remote associations buried in a memory. These kinds of hidden relationships, brought to the fore, are the stuff of insight.

Sleep's abilities to underscore the obscure and help us to forget unproductive lines of thought are not different gifts. They are, in fact, two sides of the same coin. If you are stuck in a rut while working on a problem, it's probably because you can't shift your attention away from a conspicuous, but futile, line of thought: *Fire is a danger; pliers a tool*. This makes you less able to consider more subtle and fruitful ideas: *Fight fire with fire; use the pliers as a pendulum weight.*

WAKING INCUBATION

If incubation during sleep were easy to study, then we would know much more about it. Fortunately, incubation can also take place while you're awake. This is much simpler to investigate, which is why most incubation research has focused on alert participants.

Waking incubation is actually quite common, even if it often goes unrecognized. For example, once Mark was waiting in the checkout line of a grocery store. The clerk was cleaning the counter when the nozzle of her spray bottle came off, leaving the plastic tube stuck deep inside the bottle. She tried to reattach the nozzle, but the tube was too far down. After a minute or two of fussing, she turned to greet the next customer and commenced ringing up the purchase. She had scanned a few items when her head suddenly popped up. "I know! I'll jam a pen inside the tube and pull it out!" She smiled, shook her head, and added, "I always come up with these great ideas." Many people find such practical examples of incubation more convincing than stories of great sleep-inspired discoveries or inventions. Almost everyone can recount similarly humble anecdotes. However, such stories can't prove that waking incubation is beneficial, because they don't tell us whether a particular problem would have been solved just as quickly without diverting attention. To discover whether waking incubation helps, we have to directly compare it to uninterrupted work on a problem.

Many such incubation studies have been done. However, until recently, this body of research has been inconclusive. Some studies have shown incubation to increase solution rates compared with uninterrupted work. Others have yielded no evidence for such a boost, leading some researchers to suspect that the benefits of incubation may be more of a myth than a mystery.

The inconsistency of past incubation research is likely due to the variety of methods used to study it. Some experiments used short incubation periods, while some used long ones. Some used verbal problems, some visual ones. The incubation periods of some experiments were filled with difficult distractor tasks, while others used easy tasks, no task, or sleep as a time filler. And so forth. The point is that there are many ways to do an incubation experiment, and it's likely that some of these possibilities yield little or no incubation, thereby blurring the picture. But just because it's possible to design an experiment that shows no incubation doesn't mean that incubation, waking or otherwise, is a myth. It just means that incubation isn't equally effective in all kinds of situations.

More precision was needed, and two psychological scientists, Ut Na Sio and Thomas Ormerod, recently put the question to rest. They took the findings of all available prior studies of incubation and performed a statistical "meta-analysis" that essentially combined them all into one giant high-resolution study. Their central conclusion was that incubation is real—if a person takes a break from a problem and returns to it later, then the break can increase the likelihood that she will solve the problem (relative to uninterrupted work).

FATIGUED BRAINS, AUTOMATIC THINKING, AND THE REAL UNCONSCIOUS MIND

Mental work can be draining. Even a few minutes of intense concentration can sap what cognitive psychologists call the brain's "executive functions." These executive functions enable you to set goals, suppress irrelevant thoughts, focus your attention, control your memory, and so forth—all things that you need to do to solve a problem. Common sense suggests, and recent research verifies, that a period of rest should allow you to return to a difficult problem with renewed mental vigor to improve your chances of achieving a solution.

However, the mental-fatigue hypothesis doesn't explain some of the most important features of incubation. If during your break you perform a task that continues to demand mental effort, then the break can still help you to solve the problem. And mental rest certainly doesn't explain examples such as those of Otto Loewi and the grocery store clerk who took a break only to have the solution pop into their minds while they weren't actively working on the problem. The fatigue idea implies that rest should help you solve a problem only when you resume working on it and not before. Clearly, a break can have benefits utterly distinct from those accruing from rest.

Some researchers have proposed "unconscious thought" as the

mechanism for incubation. Their version of unconscious thought is not the same as the unconscious incubation process we are discussing here. According to their idea, when your conscious mind takes a break from a problem, your unconscious mind—which these researchers contend thinks just like your conscious mind—picks up the slack and continues to work on it. Then, when your unconscious mind solves the problem, it conveniently sends a message to your conscious mind in the form of an epiphany.

The unconscious-thought idea is attractive because it appeals to our lazy side. Isn't it much better to let your tireless servant chug away on a problem so that you don't have to invest the time and energy? Got a tough problem to solve? Go play golf and let your faithful brain servant work on it while you relax and have fun.

Alas, depending on exactly what you mean by "unconscious thought," there is no consensus among scientists that such a thing exists. Some proponents of unconscious thought contend that it's a particular type of complex problem solving—essentially, analytic-style thought—continuously taking place outside of awareness. Here's an analogy that illustrates this idea: You have a day-shift job that requires you to solve puzzles, say, anagrams. You're working on solving a particularly difficult anagram by methodically moving around the letters to find the combination that makes a word. It's five p.m., which is the end of your shift, and you haven't yet solved the puzzle. Your boss doesn't want to pay you overtime, so he tells you to quit for the day. You hand off the problem to a colleague who continues to work on it during the night shift in exactly the same way you did during your day shift—by methodically trying out letter combinations. Meanwhile, you go home and cook dinner. At some point during the evening, he solves the puzzle and surprises you by calling you at home to tell you the solution.

In this analogy, you are your conscious mind and your coworker is your unconscious mind. He worked on it without a break in exactly the same way you did during your shift, except that you weren't aware of what he was doing because you were home cooking. And when he called you with the solution, this was your "insight."

Some recent studies claim to have found evidence of this kind of continuous unconscious analytic thought. In our view, these studies generally haven't employed the most rigorous research methods and haven't been replicated convincingly, so we don't think at this time that there is any compelling evidence that you have a hidden "mini-me" who thinks just like you do and who will work for you tirelessly and without complaint whenever you take a break. But a massive amount of evidence confirms that the human brain can process information outside of awareness in other ways.

One of these has nothing to do with incubation. It's like an automatic reflex, akin to your knee jerking when the doctor taps it with a little mallet. For example, when you will yourself to walk across the room, this triggers your brain to perform complicated, preprogrammed computations involving the analysis of your environment, the planning of muscle movements, and so forth. Through repetition, your brain has automated these computations, so they become unconscious and effortless. But this mechanization is achieved only through practice in situations that don't change much. If you are right-handed and try writing with your left hand, you will quickly realize that such automated processes break down when a situation changes just a bit. Then you have to do things consciously and deliberately.

These automated processes can't be thought of as "creative" because they don't have the flexibility to deal with novel situations. They can, however, become tools in the service of creativity, as when a jazz pianist draws on his musical technique to help implement a novel improvisation.

Another type of unconscious cognition is what makes incubation possible. It operates like a tour guide who points out and explains important landmarks. This guide doesn't think for you but does provide a perspective for understanding what you see. Now, imagine what would happen if every monument you visited had its own tour guide, each of whom had a unique view: historical, architectural, religious, or philosophical. Every time you went from one monument to another, you would experience a change of perspective. Each shift of perspective enables you to think about these monuments in a new way. This is how incubation leads to insight.

To understand this in scientific terms, consider that every piece of information stored in your brain has associative connections to other pieces of information. "Dog" is connected to "bark" and "cat"; "Big Mac" to "fries" and "McDonald's"; "movie" to "ticket" and "popcorn"; and so forth. When you let your mind wander, these associations bubble up into awareness as a loose sequence of thoughts—the stream of consciousness, in which each idea evokes the next idea in a chain. Some associations are generally stronger than others: The word "dog" would ordinarily be more likely to elicit the word "cat" from you rather than the word "battery." However, influenced by your ongoing thoughts and environment, individual associations wax and wane in strength. For example, "dog" might suggest "catcher" or "pound" if you're walking down the street, "hot" if you're attending a baseball game and are hungry, "veterinarian" if you're shopping at a pharmacy, or "battery" if you're in a toy store.

So how you think about something is determined not only by your past experiences and habits but also by the current state of your vast network of associations. These associations are like a pair of glasses that filter your perceptions of the world around you. They color what you see and make it easier for you to focus on some things rather than on others. But the state of your associations is malleable and constantly in flux. Every time your circumstances change, so do your glasses. That's how a change in context can help you see things in new ways. This view of the unconscious mind as a filter or spin doctor for the conscious mind, rather than as a hidden collaborator or servant, provides a plausible basis for theories of incubation.

UNFINISHED BUSINESS

During the 1920s, Soviet psychologist Bluma Zeigarnik traveled from the Soviet Union to Germany to study psychology with the Gestalt psychologist Kurt Lewin at the University of Berlin. Lewin once recounted an informal observation to her. During a meal in a restaurant, he was impressed by the waiters' ability to remember multiple complicated orders without writing them down. But when Lewin directly questioned the waiters, his illusion about their memory ability was shattered—as soon as a waiter finished serving the customers at a table, he forgot their orders. This made a deep impression on Lewin, but it made an even deeper impression on Zeigarnik, whose subsequent research on this phenomenon culminated in a famous paper she published in 1927. She demonstrated under controlled laboratory conditions what is now known as the "Zeigarnik effect": An interruption improves memory for the unfinished task; but once completed, the odds of forgetting the task suddenly increase.

More recently, cognitive psychologists Colleen Seifert, David E. Meyer, and their colleagues went a step further: When you tackle a problem and fail to solve it, it sticks in your craw—and your brain. Seifert not only showed that you are more likely to remember problems that stumped you than problems that you were able to solve, but also proposed that the memory of an unsolved problem carries within it the seeds of its own solution. Their idea is that the failure to solve a problem stimulates your brain to store a special, easily retrieved memory of the problem and the failure. This memory is much more than a mental note. It energizes all of your associations to the information in the problem, sensitizing you to anything in your environment that might be relevant, potentially including the solution. Thus, when you encounter something that's even remotely associated to the problem—a word, a sound, a smell—it can act like a hint that triggers an insight. To use their term, you "opportunistically assimilate" the hint. This energizes a weakly activated, subconscious idea so that it can emerge into consciousness as an insight.

In 1995, on the puzzle segment of the National Public Radio program *Weekend Edition,* the puzzle master asked listeners to take the two words "shout" and "danger" and rearrange the eleven letters to form two words that are antonyms. Edward Bowden, one of our research collaborators, heard the program and gave the anagram a lot of thought, but he was stumped. The solution came to him later while attending an opera, Mozart's *Don Giovanni*.

Bowden had his aha moment during Act I, when Don Giovanni fatally stabs the Commendatore while the latter was attempting to defend his daughter's honor. The timing of Bowden's revelation surprised him because he wasn't actively thinking about the puzzle at the time. But as an insight researcher, he couldn't let this kind of mystery just pass, so he thought about why the solution might have occurred to him when it did. Then he realized what had happened. The English translation of the opera's libretto, projected above the stage, contained the word "daughter." When Bowden saw this word he suddenly realized that the puzzle's solution was "son" and "daughter." He wasn't consciously thinking about the puzzle, but reading the word "daughter" sparked his insight. However, by itself, simply reading the word "daughter" may not have been enough to trigger the solution. Seeing the Commendatore angrily confront Don Giovanni and knowing what was about to happen likely primed the puzzle words "shout" and "danger." This sensitized him to the hint "daughter" when it was projected above the stage. Seeing "daughter" then triggered the complete solution.

Bowden followed up on his operatic experience with a laboratory study. He asked participants to solve a series of remote associates problems. Just before each problem, he flashed a hint word on the screen so briefly that they couldn't consciously identify it. Even so, their brains were able to process these words unconsciously. When these subliminal hints were related to the solutions, they helped people solve the problems. Importantly, the solutions sparked by the hints were usually insights rather than products of analytic thought.

So, the failure to solve a problem can sensitize you to hints around you, even when you aren't aware of these hints. When this happens, the solution is likely to pop into your awareness rather than stimulate you to consciously analyze your way through the problem. This has to make a person wonder how often things around us—things that we aren't paying attention to—trigger our insights. There are obvious benefits to this mechanism. Unfortunately, these benefits carry with them a risk: unconscious plagiarism.

Imagine that a person heard or read a great idea and then promptly forgot it. If he retained this idea subconsciously and it later popped into awareness when triggered by a subtle cue in the environment, then it might seem to him like an original creative insight because he couldn't consciously trace back the logical steps that produced it. His implicit logic might be "If it came to me all of a sudden, then it must be my idea!" Sorry, not always. Many a scientist has had an "epiphany" of some sort, only to find out later that someone had already pitched the idea to him during a hurried conversation.

Opportunistic assimilation probably explains many incidents of insight outside the laboratory. Its limits haven't yet been established, but there is clearly some practical advice here. When you take a break from a problem, you'll increase the odds that you will encounter an insight trigger if you expose yourself to a variety of stimuli and environments during your break. For example, walking tours often did the trick for nineteenth-century British poet William Wordsworth. Walks through France and North Wales as well as his English homeland, especially the Lake District, inspired him to write. Dorothy, his sister, often accompanied him on these walks and explained in her own *The Grasmere Journals* that various sensations and views during those strolls sparked both siblings' creations. She noted that a "beautiful" London morning on July 27, 1802, was an opportune time to cross the Westminster Bridge as they experienced what later led her brother to write the meditative, uplifting sonnet "Composed Upon Westminster Bridge." She wrote, "The houses were not overhung by their cloud of smoke & they were spread out endlessly, yet the sun shone so brightly with such a pure light that there was even something like the purity of one of nature's own grand Spectacles."

Another practical benefit of hints is that they can sometimes trigger the solution to a problem that you didn't even know you had. Spontaneous insights can occur when your unconscious associations are tweaked until something deeply buried, or even entirely new, pops out.

Richard James was an engineer aboard a U.S. Navy ship in Philadelphia in 1943 when he was assigned the task of figuring out how to stabilize the instruments on his ship so that the instruments' displays would be easier to read. He tried using springs as shock absorbers to cushion the instruments against the ship's movements. While he was working on one of these instruments, a spring popped out and bounced around the room as if it had a life of its own. James suddenly realized that such a spring could be the basis for a toy. It took him a couple of years to perfect his creation, the Slinky.

Experiencing variety not only exposes you to insight triggers, it can also suppress insight blockers.

GET OVER IT!

Erik Verlinde was vacationing in the South of France during the summer of 2009 when a burglar stole his passport. "I had to stay a week longer," Verlinde later said. Then, "I got this idea."

During that extra week, Verlinde, a renowned University of Amsterdam theoretical physicist, sent a series of email messages to his twin brother, Herman, an equally distinguished Princeton theoretical physicist. The first message said that he had been robbed and had to delay his departure. The next one described his new idea. Subsequent messages fleshed out its underlying mathematics.

Erik Verlinde's insight was that gravity doesn't exist.

"When this idea came to me, I was really excited and euphoric even," Erik said.

"What's going on here? What has he been drinking?" Herman thought.

Erik doesn't deny that things fall—just that they fall either because objects exert a pulling force (Newton) or warp the space around them (Einstein). He argues that gravity doesn't exist as a separate thing. The underlying mathematics is very advanced—a type of algebra was named after him—but the gist is that instead of being a fundamental force or property of the universe, gravity is a by-product or emergent property of other, more basic, processes, just as a baseball game results from all the individual actions of many players. The universe tends toward disorder. Objects fall because falling is the most disorderly and likely outcome to emerge in this complex system. The idea resulted in a groundbreaking and controversial paper that aims to revolutionize our understanding of the universe. Physicists are currently evaluating his theory.

For psychologists and neuroscientists, the most interesting thing about Erik's idea is that it didn't come to him while he was sitting in his office at the University of Amsterdam or while he was engaged in a deep discussion with colleagues at a physics conference. Rather, his vacation and the disruptive theft seem to have nudged Verlinde's brain into a state conducive to insight. "It's interesting how having to change plans can lead to different thoughts," his brother said.

So, fixation forgetting isn't confined to sleep. If you get stuck while working on a problem, then focusing on something other than the problem can allow the unproductive idea to dissipate and less obvious possibilities to emerge.

Of the various theories of waking incubation, fixation forgetting has the most empirical support. One notable example was a study that cognitive psychologists Edward Vul and Harold Pashler conducted over the Internet. Their participants worked on groups of anagrams or remote associates problems. Not surprisingly, when they saw a misleading hint next to a problem, they were less able to find the solution. But taking a break from the problem reduced distraction from the misleading hint because it loosened the hint's grip on their thoughts. Without a misleading hint, a break didn't help. And without a break, a misleading hint's effect persisted because there was nothing to stop it from dominating thought. A bad hint is like a bad guest—it lingers and monopolizes the host until a good guest cuts in to rescue him.

Being a good guest requires impeccable timing. If she cuts in and displaces the bad guest immediately, then she would be considered rude. But if she waits too long, then the host suffers. Timing is everything. The best time to cut in is just when the bad guest starts to become annoying.

Cognitive psychologist Jarrod Moss and colleagues gave helpful hints to participants either before they got stuck, right at the point of impasse, or some time after. They found that these hints were most effective when presented right at the point of impasse. If a helpful hint is presented too early, there is no benefit because the solver hasn't yet worked on the problem long enough to fully grasp it. On the other hand, if the hint is presented too long after an impasse is reached, then the unproductive line of thought has become too entrenched, making it difficult to dislodge. It's easiest to pluck a bad idea before its roots grow too deep.

So there is strong evidence that a break can help you to get over a wrong perspective. Yet there is even more to this idea. What you think about during your break makes a difference.

To get the maximum benefit from a break, it helps if you "change your mind"—the greater the change, the better. Cognitive psychologist Peter Delaney and colleagues asked participants to learn a list of words and then daydream about an assigned topic until their memory for the words was tested. The topics were either more or less related to the participants' daily lives. People who were told to daydream about their parents' house remembered more words than when they were told to daydream about their own house. People who daydreamed about a vacation in a foreign country remembered more words than if they daydreamed about a domestic vacation. So thinking about situations that are different from your current daily experiences can help to purge your mind of sticky thoughts. Out with the old; in with the new.

Context is the key. Your memories are associated with the contexts in which you originally acquired them. If you learn something in one place, type of situation, or frame of mind, then you'll remember that information better if you recall it later in the same context. That's why you sometimes can't remember the name of a person you know only from your office when you encounter that person at the shopping mall. (Cognitive psychologists call this the "butcher-on-the-bus phenomenon.") So when you're stuck in a rut and take a break to resolve the impasse, you'll shed the misleading ideas more quickly and thoroughly if you think about things that transport your mind far from your current situation. Of course, actually getting away and taking a walk, especially on an unfamiliar route, would be an even better way to change context (as long as you aren't thinking about the problem during your walk). But if you can't get away, try *thinking* about getting away, to another town, another country, or, even better, another planet.

Thinking about a situation far removed from your current one might have additional benefits beyond those afforded by a mental reset. When you do this, you are also exercising your *ability* to change perspective, which seems to be the same sort of ability that you use when you reinterpret a situation to achieve insight. Later, we'll tell you about research that suggests that exercising this mental flexibility can, at least temporarily, boost your insightfulness.

STATE OF THE ART

To sum up, failing to solve a problem sensitizes you to things in the environment that may subsequently trigger an insight. Taking a break when you are stuck or fixated can purge the wrongheaded perspective that traps you in a mental rut. Exposing yourself to a variety of experiences or thoughts—especially unusual ones—during a break will help you to dismiss these unproductive thoughts and increase your chances of encountering an insight trigger. The legendary late jazz pianist and composer Erroll Garner once noted, "I get ideas from everything. A big color, the sound of water and wind, or a flash of something cool."

As for the "unconscious thought" idea, we haven't yet seen any convincing evidence that an awake person has an invisible alter ego in the brain that can solve problems in the same way that you consciously and deliberately solve them. This idea can't yet be completely ruled out. But for now, the best evidence indicates that incubation leads to shifts of perspective by tuning the strengths of the unconscious associations that filter our conscious perceptions. That's why you should work on a problem until you solve it or get stuck. Only then should you take a break to cleanse your mental palette. And if you fall asleep during your break, then that can supercharge your ability to dismiss wrongheaded ideas and discover hidden relationships. But there may be more.

It's likely that the limited amount of rigorous science hasn't yet caught up with everyday experience regarding incubation during sleep. There are so many documented examples of powerful creative breakthroughs occurring during or immediately after sleep that we have to take seriously the notion that sophisticated analytic thought can solve complex problems while you slumber, even if this kind of unconscious problem solving doesn't take place while you are awake. Our receptivity to this idea was recently bolstered by a student's personal experience.

Jason van Steenburgh, at the time one of John's graduate students, had run into a problem in the lab. He had already spent several months performing a complicated EEG study (unrelated to insight or problem solving), after which he began analyzing the data. Though the EEGs of the first seven participants looked fine, he was horrified to discover that the brain waves from the remaining thirty-five participants seemed to be random squiggles. Something was very wrong.

Jason's fear was that an undetected technical problem had caused him to waste several months collecting junk data. He spent the next three days poring over every aspect of the experiment to try to find the problem, but everything seemed to be working properly. He couldn't think of anything else to check. Months of work, lost. Despondent, he resigned himself to repeating the experiment.

That night, Sasha, Jason's two-year-old daughter, appeared to him in a dream. They were in the EEG lab. She said, "Daddy, it's the PPM files." Like Otto Loewi, Jason quickly forgot the dream after waking in the morning. However, later, while feeding Sasha, Jason suddenly remembered the dream. He promptly went to the lab to check his "dream daughter's" idea.

Every computer file with EEG data had an accompanying "PPM" file that was needed to analyze the data. Jason checked these PPM files and discovered that the thirty-five participants' worth of junk data had each been analyzed with the wrong PPM file. That's why the results were garbage. Fortunately, it wasn't too late to go back and repair the damage. After reprocessing the data files, he found that they were all usable.

Thanks, Sasha, for inspiring your father's insightful dream!

So take a break. Even better, get plenty of sleep. And dream. Perhaps it will help you to solve a difficult problem. If so, then write down the solution immediately! But if the extra sleep doesn't help you to generate any new ideas, then at least it will make you feel more refreshed and in a better mood, which just happens to be the focus of the next chapter.

9

IN THE MOOD

Necessity is the mother of invention—my necessity was to entertain my kids.

—Jerome Swartz, co-founder, Symbol Technologies, Inc.

Jerry Swartz is an accomplished engineer and prolific inventor with more than two hundred patents. He is also an innovative entrepreneur, having founded and led a major company, Symbol Technologies, which was acquired by Motorola for $3.9 billion in 2007. But his most famous idea didn't occur to him at a board meeting or during discussions with his engineering staff. It didn't even happen at work.

Whenever Jerry Swartz came home after a day at the office, his young children would want to know what interesting things had happened. One evening in 1980, he brought home a laser pointer to entertain them. In those days, laser pointers weren't the miniaturized handheld devices that people now use to give presentations; they were housed in bulky, foot-long metal tubes. But they were cutting-edge and fun. He delighted his kids by pointing its beam through his living room window and tagging passing cars. Then the reverie was punctuated by an insight: "I moved the thing on the side of the car. . . . As I moved my hand up and back, it was able to stay with the car, pretty much in one spot. So the idea occurred to me . . . about moving a spot of light, and how a scanner works."

This insight triggered a chain reaction of additional thoughts about how to flesh out this idea: "I started thinking . . . my wrist motion is doing it, so why can't a motor do it? . . . What if I embedded the laser itself in the barrel of a gun and do something not to move the laser tube . . . but just move the beam of light? For instance, bounce the laser off a mirror . . . mounted on a little stepper motor . . . moving the beam in a line."

That was Swartz's aha moment. Within a year, his company started selling the first handheld laser bar code scanner. They are used in many commercial applications, but most people have seen one particular device based on Swartz's idea: the pistol-like scanners used to ring up items in most nonfood retail stores in the United States. Within a couple of years, his company was selling millions of them.

Some insights seem to come completely out of the blue. Swartz's didn't. It was the culmination of a lengthy period of immersion and incubation. Laser bar code scanning technology already existed in 1980. Indeed, a pair of Drexel University graduate students first conceived of bar codes and scanners in 1948, for which they were awarded the first bar code patent in 1952. The first commercial use of bar code scanning occurred in 1966, and the first retail bar code scanner was used in 1974 to scan a pack of Wrigley's Juicy Fruit gum in a Marsh supermarket in Troy, Ohio. But until 1980, all the scanners were large devices built into checkout counters. As anyone who has hoisted a fifty-pound bag of dog food onto one of these counters knows, a portable device has advantages.

Swartz had previously considered the technical feasibility of a portable scanner and had discussed it with others in the industry. Existing countertop scanners used unwieldy metal-encased lasers to project complex moving polygons across a bar code. This kind of system couldn't be easily miniaturized. Swartz had "been trying to think of very fancy ways to do it . . . trying to model the mathematics behind the system design of the lenses, the lasers, the spot of light coming out." Nevertheless, "it still hadn't quite come together" for Swartz until that evening in 1980.

Why then? He wasn't thinking of scanners that evening, only of amusing his kids. They were simply having fun, laughing and relaxing. They were in a good mood.

Swartz's story is not unique. For example, Alexander Graham Bell had spent years trying to invent the telephone, but all his preliminary ideas proved infeasible. Then one day he went to a favorite place in the woods overlooking a river, sat in a wicker chair, and relaxed as he watched the currents of water flow by. While enjoying the natural beauty that surrounded him, he suddenly realized that sound waves could be transformed into flowing currents of electricity. Conducted along wires, these electrical currents could be converted back into sound waves at a distant location. This idea was the basis for his invention of the telephone.

A common thread runs through the stories of Jerry Swartz, Alexander Graham Bell, and many other creative figures. Hermann von Helmholtz explained how a good mood and relaxing walks in the country stoked his creativity. Art Fry was happily singing in church, fumbling with bookmarks in his hymnal, when he suddenly thought of the perfect use for a weak adhesive recently developed at his company, 3M: Post-it Notes. Judah Folkman was peacefully enjoying synagogue services when he was struck by a realization about tumors and angiogenesis. A quote attributed to Wolfgang Amadeus Mozart, "When I am, as it were, completely myself, entirely alone, and of good cheer—say, traveling in a carriage or walking after a good meal . . . it is on such occasions that my ideas flow best and most abundantly," and other such anecdotes would seem to suggest that a positive mood enhances creative insight.

FROM THE LAB TO THE WORKPLACE

Perhaps informed by such stories, some corners of corporate America—Google, Pixar, IDEO—strongly believe that environments that encourage contentment, even play, foster outside-the-box thinking and innovation. In contrast, other corporations try to encourage innovation by cranking up the pressure cooker. Support for both approaches has been based largely on an accumulation of anecdotes, assumptions, and plain guesswork. Let's move beyond the subjective to examine what we actually know about how mood influences creative insight.

Beginning in the early 1980s, social psychologist Alice Isen and colleagues conducted groundbreaking research examining the effects of both positive and negative moods on people's ability to solve problems. She showed that when people are in a state of calm happiness, they are better able to solve problems that usually require creative insight rather than analytic thought. It didn't matter how a positive mood is induced: whether by giving people unexpected gifts or praise, having them recall happy experiences, having them watch comedy films, or having them listen to upbeat music.

Some of Isen's laboratory investigations ventured into more complex forms of problem solving to show, for example, that physicians display more flexible diagnostic reasoning while in a positive mood. Nevertheless, some people might ask whether her laboratory studies shed light on their daily lives. Do these findings illuminate the effects of mood on creativity in real-world environments such as the workplace?

Yes. Social psychologist Teresa Amabile launched an ambitious project in 1999 to track workplace insights and identify their emotional antecedents. Her team trained 222 employees from seven companies in three different industries (chemicals, high tech, and consumer products) to make daily diary entries and respond to electronic prompts. In the end, the researchers collected more than eleven thousand questionnaires and diary entries describing sixty-three thousand work and life events. In their narratives, the diarists were asked to "briefly describe one event from today that stands out in your mind as relevant to the target project" and later to "add anything else you would like to report today." The respondents mentioned 364 breakthrough ideas, and Amabile's team compiled a wealth of other information, including information about participants' self-reported and peer-reported moods.

When the researchers analyzed all of these data they found an astonishingly strong association between increased happiness on certain days and important breakthroughs at work on the following day or two. The boost that a positive mood gave to creative thinking lasted for at least two days, apparently affecting the incubation period leading up to a solution. (Of course, solving a problem also made them happy, though usually not for longer than a day.)

Amabile's workplace findings generally confirmed those from laboratory studies, with one difference. Isen's lab studies showed that enhanced creativity occurred during a positive mood, whereas Amabile's workplace study showed enhanced creativity before and shortly after a period of positive mood. This minor discrepancy was likely due to differences in the time scale of these studies. A participant in an experiment is usually in and out of a lab in an hour or less, and the changes in mood are minor and fleeting. The workplace study included slower, stronger, real-life emotions and looked at problems that took days or weeks to solve. Perhaps the slower, stronger positive moods of the workplace energize the incubation that precedes an aha. Or perhaps these more intense emotions must cool off a bit before a complex workplace problem can be solved with an insight. Either way, it's clear that a positive mood facilitates creative insight in everyday situations as well as in the laboratory.

CROSSTALK

But it cuts both ways. Emotions influence how you think, and how you think can influence your emotions. Many people incorrectly assume that the dialogue between thought and emotion occurs between two separate, different, and unequal partners, at best requiring a translator, at worst an arbitrator. And some are suspicious when this conversation becomes a little too intimate: Those whose thinking is too emotional are often regarded as lacking objectivity or rationality; those whose emotional experiences contain too much logic are likened to cold, unfeeling computers.

Until fairly recently, these assumptions have rarely been questioned. Many cognitive psychologists and neuroscientists take pains to minimize the "contaminating" effects of emotion in their experiments, while emotion researchers often pay no more than lip service to the role of thought in emotion. But the fact is that cognition and emotion are closely intertwined, even blended, in the brain.

This blending can take several forms. One of the best known is "mood congruency": If you're sad, then you'll tend to think of sad things, even to the point of interpreting ambiguous situations negatively. Likewise, happy people tend to have positive thoughts and interpretations. Just as our unconscious associations are filters through which we view the world and ourselves, so, too, are our moods. Indeed, a major component of psychotherapy for depression is the attempt to eliminate inaccurate or maladaptive filters, both cognitive and emotional, so that people can see things the way they really are.

But mood congruency doesn't explain why positive mood enhances creative thought. There's nothing inherently positive about bar code scanners or telephone communication. For sure, innovation can lead to tangible rewards. But as we'll see later, when people know that solving a problem can lead to a specific payoff, a breakthrough solution may actually be less likely to occur. So mood must be changing something else in our thinking and in our brains.

FOCUS

One afternoon many years ago, John was walking down a city street when he saw two women struggling just off the curb. At first, the sight startled him. Then he realized that one was roughing up the other, so he decided to try to break up the fight. But when he was about four or five feet away, he saw that one of the women had a gun stuck in the other one's ribs and was trying to yank away her victim's purse. Seeing John approach, the assailant turned and pointed the gun at him. Naturally, he backed off. While doing this, he thought that he would study the assailant's face carefully so that he would be able to give the police an accurate description of her. He tried hard to do this, but couldn't. His attention was locked on the gun, and he didn't have the strength of will to shift or expand his focus. The perpetrator grabbed the purse and ran away, and later John was able to give only a vague description of the thief. But he was able to give a detailed description of the gun. (Thankfully, the police picked up the perpetrator several hours later driving the victim's car; she had stolen the purse to get the car key.)

This is an example of "weapons focus." Visual attention works something like a spotlight, highlighting a part of a scene while leaving the rest obscured in darkness. Anxiety causes tunnel vision, making it easier to focus on a threatening stimulus such as a weapon. Even mild anxiety narrows and brightens the spotlight.

From an evolutionary perspective, this is logical. An early human spotting a lion in the distance while wandering on the savanna would want to keep him in keen focus and not be distracted by anything less dangerous. In the modern world, a modest dose of anxiety can help a student focus on preparing for an exam or it can help a manager persist in seeking a solution to an office problem. A little anxiety can sometimes even help people solve problems—when analytic thought suffices. But if there is useful information outside of the spotlight of attention, then this narrowness comes at a cost. Focusing on a gun will lead you to the wrong conclusion if you can't broaden your attention enough to notice that it's a toy being carried by a child wearing a Halloween costume.

Now look at the following two figures:

Which one of those two figures is more like the next one? Take a moment to think about this.

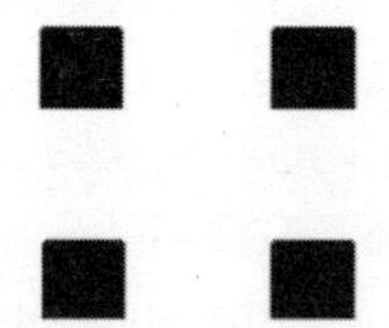

This test is used in laboratory experiments to assess how you are focusing your attention. There is no correct answer. If you're attending broadly, you'll think more about the overall pattern and say that the square made of triangles is more like the square made of squares. But if your attention is narrowly focused, you'll tend to notice the parts rather than the whole and think that the triangle made of squares is more like the square made of squares.

Emotion researchers Karen Gasper and Gerald Clore asked people to recall happy or sad personal events to put them in a positive or negative mood. Then they had them view a series of figures like these. Those people in a positive mood based their judgments on the overall pattern; those in a negative mood based their judgments on the parts. It seems that a happy mind is free to roam the forest, while an anxious mind prefers to hide in a tree.

Practically speaking, it's easy to see how a positive mood could help, for instance, an architect, by making it easier for him to envision large-scale forms, like buildings. But many types of problems aren't visual. It's not immediately obvious how broadened attention could help a person solve an abstract or a verbal problem in a creative fashion.

ILLUMINATING IDEAS

It just so happens that the spotlight of attention can illuminate even intangible ideas. Thinking, as well as seeing, can be narrow or broadly inclusive. Some of the same brain processes that regulate the breadth of visual attention also regulate the scope of a person's "conceptual attention," the ability to focus on ideas. Both perception and thought are narrowed by a negative mood and broadened by a positive mood.

Cognitive psychologist Adam Anderson and colleagues tested this idea. They measured the breadth of people's perceptual attention by having them identify a target letter in the center of a display while ignoring the distracting letters that surrounded it. This is similar to what happens when you are driving and see a red stoplight ahead but can also see another lane's green light. That green light is an interfering "flanker" stimulus. It takes a fraction of a second to suppress the urge to continue driving when you see it. If you were a Zen master with extreme control over your attention, you might be able to focus so narrowly on your red stoplight that you could completely ignore the other lane's green light and not experience the interfering tendency to drive on. Some people are better at this than others, but few of us are Zen masters, so almost everyone responds more slowly when flankers conflict with the central target.

Anderson and his team found that a positive mood broadens people's perceptual attention, making it harder to ignore flankers. (This suggests that happy drivers are more likely to drive through a red light when a green light is next to it!) Furthermore, a negative mood makes it easier for the participants to focus and exclude the flankers.

The researchers then asked whether mood has the same effect on conceptual attention. Broad conceptual attention helps a person solve remote associates problems because the solutions are distantly related to the problems. That's because when your scope is wide, you are better able to come up with the loose associations you need.

Happy participants solved more remote associates problems. They also showed broader visual attention in the flanker task. Those in a negative mood showed the opposite pattern.

Thus, your mood influences your scope in a general way. Seeing the big picture means thinking it, too. Happy people think with distant associations and inclusive categories, which is why they tend to see connections and commonalities among things. For them, everything seems relevant. Everything is One. For those in a foul mood, things seem different, separate, and unconnected. They split hairs. Of course, sometimes noting fine-grained distinctions is what you need to do. But when carried to extremes, the result can be a narrowness of both the senses and the mind.

All of this is quick and unconscious, almost reflexive. Recall that in Chapter 3 we discussed the N400, a fast EEG brain response signaling that a word or picture doesn't fit into its context. If a person reads or hears the sentence "He puts cream and sugar in his dog," the word "dog" will elicit a whopping N400 response compared with the expected word, "coffee," because "dog" doesn't fit. But in "He puts cream and sugar in his cocoa," the word "cocoa" is a little off, but it isn't terribly inappropriate. It sparks a medium-sized N400 somewhere in between those from the expected "coffee" and the outrageous "dog."

But all this depends on how you are feeling. When you're in a buoyant mood, "He puts cream and sugar in his cocoa" doesn't seem all that strange or unexpected, so "cocoa" doesn't raise much of an alarm. A happy person's N400 response to "cocoa" is, therefore, small. But when you're in a bad mood, "cocoa" seems unrelated to coffee and feels like an outrageous way to conclude the sentence, so "cocoa" prompts a large N400 protest.

As we've seen, the brain is constantly trying to predict what will happen next, emitting one or another type of fast EEG bursts whenever its expectations aren't met (Chapter 3). These bursts are alarm bells warning that the brain may have to replace an inadequate mental box. However, these boxes aren't static structures that must either be kept or shattered. They are adaptable. A positive mood can inflate a box to make it more expansive and accommodating, affording possibilities for flexible, far-ranging thought. A negative mood can shrink a box to make it less inclusive and more sensitive to its borders—the "Get off my lawn!" effect.

Positive mood can have even wider effects. By broadening attention, happiness can help a person widen and integrate his or her knowledge of the world. Think of this as the flip side of weapons focus. Positive emotions such as tranquillity, joy, or love expand a person's horizons by making things seem relevant and connected. This enables him or her to set aside habitual responses, explore the environment, and consider new opportunities and ways of thinking. Playing with your kids, enjoying the beauties of nature, listening to inspiring music, or walking through an art museum, for instance, will broaden your mind and your possibilities for interacting with the world, as such things did for Jerry Swartz, Alexander Graham Bell, and Hermann von Helmholtz.

Now let's look under the hood.

MARRYING HEAD AND HEART

A student tackles a set of remote associates problems in Mark's laboratory at Northwestern University. The problems are deceptively difficult, and the research participant correctly solves only eight of the first twenty within the time limit. Then, he watches an excerpt from a Robin Williams HBO comedy special. He laughs out loud as Williams depicts the "sadistic" invention of the game of golf. After watching the clip, the student rates how happy, anxious, and stimulated he feels. The comedy clip has greatly improved his mood. He's then given another set of problems and does better, solving ten of twenty. Could this improvement be a coincidence? After a third comedy clip, he solves thirteen of twenty problems, much improved from his baseline performance. Importantly, virtually all of this 63 percent improvement was due to an increase in the number of solutions by insight, with no increase in the number of analytic solutions. This suggests that being put in a good mood enhanced his creative thinking but not his analytic thinking.

But it's important to consider other possible explanations. Perhaps the comedy film had simply made him more alert. He is, after all, a college student, and is attempting these problems on a Monday afternoon, possibly after a weekend of partying. His self-ratings show that he is slightly more alert after the comedy clips than he was at the start of the experiment. But what follows shows that his improvement can't be due to increased alertness.

He watches another film clip: a terrifying excerpt from *The Silence of the Lambs*. Unsurprisingly, he now judges that he is more anxious. He also reports that he is more alert, a bit more than after the comedy clips. But when he tackles another set of problems, he manages to solve only eight of twenty. This shows that the increase in alertness from the earlier comedy clips couldn't have caused his performance to improve. Next is a horrifying excerpt from Stanley Kubrick's *The Shining*. The participant is still highly anxious and solves nine of the twenty problems in the following set. After another clip from *The Shining,* still anxious, he solves eight of twenty problems. Though his overall performance is worse than after the comedy clips, it didn't go down from his initial baseline. He achieved most of these solutions analytically, rather than through insight.

He was typical. When in a positive mood, participants (both men and women) solved more problems, all due to an increase in the number of problems solved by insight. When anxious, they solved fewer problems overall, due to fewer insight solutions (with a slight increase in the number of analytic solutions).

Anxiety, but not sadness, decreases insights. This may seem surprising, as many assume that sadness is a sort of anti-happiness. But this isn't the case. If it were, it would be no more possible to be happy and sad at the same time than it would be possible to be tall and short. Some moods can be experienced simultaneously, as in bittersweet moments such as those that parents might feel when attending a child's high school graduation or wedding—being proud and happy for their child but sad that she is becoming independent and will be moving away. However, anxiety and happiness are incompatible.

To understand how a positive mood broadens attention and enhances creative insight, we used fMRI to measure participants' brain activity during an experiment like the one we just described. We examined brain activity just before each problem was presented. During this preparatory period, participants had nothing to do other than anticipate the appearance of the next problem.

As in our earlier study, we found that preparation for insight activated several brain areas. We also found that a positive mood activated a number of brain areas. The question was whether any of the brain regions energized by positive mood were also implicated in preparation for insight. Such areas, should they exist, would be responsible for channeling the power of positive mood to establish an insight mind-set.

In fact, only one brain area was activated by both a positive mood and preparation for insight: the anterior cingulate. This structure is the doorway to a universe of possibilities.

IRRECONCILABLE DIFFERENCES

Because the human brain is limited in both speed and power, it must have a strategy for dealing with a complex and rapidly changing environment. It does this by quickly evaluating and prioritizing the things around it, lavishing attention on things deemed important and ignoring everything else. This triage is influenced by an object's emotional value, such as the allure of a fine meal or the threatening nature of a bully. But the value of things is also interpreted through the filter of your current mood. A fine meal is less appealing if you are still full from your last meal; a bully seems less threatening if you are accompanied by your friend, a martial arts instructor. The anterior cingulate makes this happen by broadening or narrowing the scope of possibilities you are prepared to consider.

A few years ago, neuroscientists discovered that one of the anterior cingulate's jobs is to monitor the brain for conflicts, such as a tendency to perform two or more incompatible responses. The original idea was that such conflicts were simple. For example, you can't push a button to your right and a button to your left at the same time with the same finger. Whenever the anterior cingulate detects such conflicting tendencies, it signals other parts of the brain's control system to settle the dispute.

This notion of incompatible tendencies can be expanded to include more complex thoughts and behaviors, such as whether to attempt to solve a problem in one way or another: *Should I try to open that stuck spaghetti sauce jar by gripping and twisting the top even harder? Or should I pretend that the jar is a can and cut open the metal top with a can opener?* There are two ways to settle such disputes.

One is the "winner-take-all" strategy (aka "the rich get richer"). Squash or ignore the weaker, less obvious tendencies and let the strongest, most obvious one dominate. This enables a person to focus on the most straightforward path to solving a problem without detours or distractions. *Just twist the jar top harder!* This is the analytic mind-set.

The other approach is the "Robin Hood" strategy of taking from the rich and giving to the poor—let a weaker, less-obvious tendency dominate over a stronger, more obvious one. *Forget brute force—use a can opener.* This is the insight mind-set.

If the anterior cingulate isn't energized enough, then it doesn't detect nonobvious, weakly activated solution ideas in the brain. In this case, the winner-take-all strategy becomes the default. Triumph of the obvious. But if the anterior cingulate is powered up by a positive mood, then it can sense the subtle presence of alternate, creative solutions and direct the prefrontal cortex to play Robin Hood and select an underdog to win. The result is an aha moment.

And when the underdog wins, there are unique consequences.

REPERCUSSIONS OF THE CREATIVE HIGH

Elation is the typical reaction to a breakthrough idea. Albert Einstein once said that conceiving the theory of relativity was "the happiest thought of my life." Judah Folkman said that an aha moment is "a very big high." Several things contribute to this emotional boost: the pleasant surprise, the accelerated thought, and the confidence and satisfaction instilled by a solution that seems obvious in hindsight. But more is going on.

Positive emotion and associative thought are flip sides of the same coin. We've seen how a positive mood improves a person's ability to think with remote associations. Surprisingly, thinking with remote associations also improves a person's mood. This is why people crave puzzles, detective stories, and the like. These things make them happy. Creative thought can even go so far as to have therapeutic benefits, as in art, music, and writing therapies. A particularly interesting example is Freudian psychoanalysis.

Sigmund Freud's therapeutic sessions consisted mainly of "free association," in which a patient reclined and relaxed on a couch, stared at a blank ceiling, and said whatever came into his head. The blank ceiling and relaxing, nonjudgmental surroundings were intended to induce a broad, open state in which far-ranging associations could flow freely. The goal of the psychoanalyst is to help the patient trace his associations back to the emotional conflict that is thought to be the root of his psychological symptoms. According to Freud, once the patient finds this root, it can be plucked. But perhaps Freud was missing something.

Remote associative thinking and positive mood reinforce each other: Thinking with loose associations improves a patient's mood, which promotes insights about herself or her relationships with others, which further improves her mood, and so forth. But let's suppose that it's the *act* of thinking with remote associates and not the *content* of these thoughts that makes a patient feel better. In other words, it may not be so important what you are thinking about, as long as you are free-associating about something—*anything*. If so, then psychoanalysis could help even if the insights gained during therapy are irrelevant or false. Free association may contribute to a patient's improved sense of well-being even without any true self-discovery.

Of course, remote associative thinking and positive mood can't reinforce each other indefinitely. Otherwise, a person's positive mood and insightfulness would amplify each other in an infinite chain reaction. We know that this doesn't happen. Every once in a while, the cycle is broken, and this can have practical benefits.

FLUCTUATING MOODS, ALTERNATING PERSPECTIVES

We've seen that a positive mood enhances insight and vice versa. But don't highly creative people—artists and writers, for example—sometimes endure extended periods of anxiety or depression?

Yes. Psychiatrist Nancy Andreasen compiled an impressive list of creative men and women from history for whom there is evidence of bipolar disorder. She also reported higher rates of mood disorders among creative figures, for instance, in members of the Iowa Writers' Workshop, than among comparable control subjects who were not in an artistic or otherwise creative profession. The explanation for this paradox has important practical implications even for people who do not have an affective disorder.

Depression is an extreme state associated with intense analytic thinking about one's problems. Just as far-ranging associative thought can improve a person's mood and expand the mind, detail-oriented rumination can distill a person's depression and focus the mind.

Writers such as those studied by Nancy Andreasen invent characters, events, and even whole new worlds. To construct rich, consistent, and polished stories, they must obsessively ruminate on all the details and their implications. They live and breathe the characters and events that they have created for months or years at a time. This kind of rumination benefits from the focus that occurs while they are depressed. But this obsessive analysis is occasionally punctuated by an insight about the worlds that they have created—typically gained during a temporary upswing in positive mood—and these breakthroughs deliver fodder for further rumination. So the varied types of thinking that people with bipolar disorder can bring to bear at different times can enhance the quality of the final product, as long as the mood swings aren't so extreme or frequent as to be disabling.

This general principle can also work for the average person whose mood swings aren't extreme. To help you achieve a creative breakthrough, do things that make you happy. Take a walk in the park. Listen to music. But after you've had your aha moment, fun time is over. You'll need to analyze and refine your new idea, and a state of bliss won't help you to be at your critical best. In this case, a mild dose of anxiety may do the trick. To regain your critical edge, go see a scary movie, ride a roller coaster, or just watch the news—whatever makes you a little nervous. A positive mood may be the best kindling for a creative spark. But once that kindling starts burning, it takes careful work to turn that spark into a useful bonfire.

It can be hard to know when it's realistic to expect that you'll have a creative breakthrough. It may, or may not, come. How, then, can you know whether to indulge in some insight-fueling fun or whether it's better to take advantage of a somewhat anxious mood to narrowly focus on your problem and keep trying to grind through it analytically? The trick is to tell if you are ripe for a great idea.

10

YOUR BRAIN KNOWS MORE THAN YOU DO

So much of the creative process takes place in the subconscious, in an area of incomplete communication; and poetry is far from being the only discipline where such things can happen. It is in such uncharted areas of mental space where some of the deepest science has its origins.

—William W. Morgan, astronomer

Rebecca Woodings had seen many doctors, but none had been able to figure out what was making her sick. She had a severe cough, low blood pressure, leg cramps, and occasional retching. She was so weak that she couldn't even stand while waiting for a bus—she had to sit on the sidewalk. Her illness made it impossible for her to fulfill her responsibilities as an economist working for a law firm.

One day in 2009, Woodings had yet another appointment, this time with a pulmonologist. This doctor, Susan Hasselquist, didn't give her any answers, either, although she had ordered some tests.

Exhausted after the appointment, Woodings went home, ate some pizza, and went to bed. Later that night, she got up to go to the bathroom and, feeling dizzy, fell down the stairs and broke her wrist. It took all her strength to pull herself together to call 911. At the hospital, she was told that her wrist would require surgery.

For Woodings, this was a low point. But Dr. Hasselquist had a hunch.

When Dr. Hasselquist first examined Woodings, she saw a desperately sick woman. Her blood pressure was a mere 90/55, and she was so weak that she had to lie down on the examination table. Hasselquist started thinking of possible diagnoses, but nothing seemed to fit. But she had an intuition. "I knew if we just kept talking I'd figure it out," she later said.

As the conversation progressed, Dr. Hasselquist noted that Woodings had a deep tan that reminded her of something, or rather, *someone*. When the doctor questioned her about the tan, she said that she hadn't been spending much time in the sun. That's when the doctor had her aha moment. Woodings's tan reminded her of photographs of President John F. Kennedy. In many photos, Kennedy also had a deep tan that wasn't the result of time in the sun. Kennedy had Addison's disease, a deficiency of essential hormones due to deterioration of the adrenal glands. This disease gives patients a tan. Hasselquist believed that Woodings had Addison's, which afflicts only four in one hundred thousand people. Confirmation from test results came a few days after Woodings's tumble down the stairs. The diagnosis of Addison's disease explained all the symptoms except for the cough, which, ironically, was why Woodings went to a pulmonologist in the first place.

An endocrinologist prescribed medications, including replacement hormones, which were able to control the disease. Woodings got her strength back and was able to return to work, sans her fashionable tan.

Where other doctors gave up, Dr. Hasselquist persisted because her intuition told her that an insight into Woodings's illness would come if she just kept talking to the patient. Without that feeling of impending insight, Hasselquist might not have kept the conversation going until her epiphany arrived.

THOUGHTS FROM THE FRINGE

Though insights often come as a surprise, sometimes we can sense that an idea is present, lurking just below the threshold of awareness, ready to emerge. This puzzling phenomenon has a strange subjective quality. It feels like an idea is about to burst into your consciousness, almost as though you're about to sneeze. Cognitive psychologists call this experience "intuition," meaning an awareness of the *presence* of information in the unconscious mind—a new idea, solution, or perspective—without awareness of the information itself, at least until it pops into consciousness.

On October 16, 1843, the great Irish mathematician William Rowan Hamilton was walking with his wife along Dublin's Royal Canal to a meeting of the Royal Irish Academy when he was struck by an idea. In a letter to his son, he later wrote: "An undercurrent of thought was going on in my mind which gave at last a result. . . . An electric circuit seemed to close; and a spark flashed forth." His advanced mathematical idea has a relatively simple basis. Think back to how you first learned to take the square root of a number. For example, the square root of 4 is 2 because 2 × 2 = 4. But then what would be the square root of –4? There is no number that, when multiplied by itself, will yield a negative number, right? Actually, this is not the case. The square root of a negative number is called an "imaginary number," which is one component of a "complex number." If one accepts the validity of imaginary and complex numbers, then it's possible to solve equations that would otherwise be intractable. However, until the early nineteenth century, no one really understood how to think about complex numbers. Hamilton's insight was a geometrical conception of complex numbers that showed how they could be multiplied and divided, an understanding that has since become a staple of modern physics and engineering.

After his insight, Hamilton, seized by emotion, couldn't "resist the impulse—unphilosophical as it may have been—to cut with a knife on a stone of Brougham Bridge, as we passed it, the fundamental formula which contains the Solution of the Problem." Hamilton's original carving has since become obscured, but a plaque was placed on the bridge (now known as Broome Bridge) in 1958 to commemorate the event (see figure 10.1). Its inscription reads:

Here as he walked by
on the 16th of October 1843
Sir William Rowan Hamilton
in a flash of genius discovered
the fundamental formula for
quaternion multiplication
$i^2 = j^2 = k^2 = ijk = -1$
& cut it on a stone of this bridge.

FIGURE 10.1: Plaque on Broome Bridge, Dublin, commemorating Sir William Rowan Hamilton's mathematical discovery. *en.wikipedia.org/wiki/File:William_Rowan_Hamilton_Plaque_-_geograph.org.uk_-_347941.jpg*

It's noteworthy that this insight, which was sudden enough that Hamilton felt the urge to mark the specific time and place of its appearance, was preceded by the feeling of an "undercurrent of thought," which persisted until it "gave at last a result." Before his aha moment, the best that he could do was to sense the presence of this undercurrent.

Hamilton's and Hasselquist's intuitions are examples of a type of mental process that is neither fully conscious nor unconscious. This in-between type, which includes intuitions, feelings, and hunches, occupies the fringe of consciousness. Since you don't have full awareness of such thoughts, you can't really justify or explain them. You sense their presence without a clear understanding of how they got there (see figure 10.2).

INTUITION IN THE LABORATORY

The intuitions that sometimes herald insights are just as fleeting and difficult to study as the insights themselves. To be understood, intuitions have to be reliably produced in a controlled laboratory setting. Let's examine several ways in which this has been done.

When you meet a new person who is a member of a family that you know well, you often can see that she resembles other members

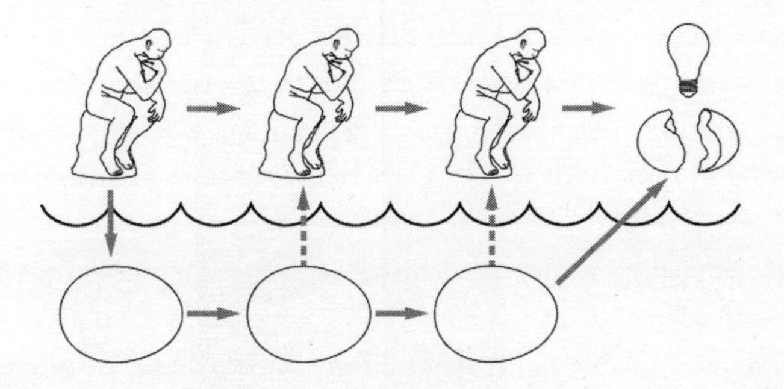

FIGURE 10.2: Intuition sometimes allows us to sense the presence of ongoing unconscious thought.

of that family, even if you can't pinpoint specific features that she has in common with her relatives. In fact, if you meet a number of strangers, you could probably do a pretty good job of intuitively judging which ones are members of the same family, even if you can't say exactly how you are able to do it.

Arthur Reber, a pioneer of intuition research, has extensively investigated these kinds of judgments. In a typical lab study, he showed participants a series of letter groups, none of which are meaningful words. One of these letter groups might look like:

RNIWKQ

In the simplest version of his experiment, each letter group was generated by a computer program according to one of two sets of rules, which we'll call "rule set A" and "rule set B." These rules govern which specific letters in a group can follow other specific letters. For example, in one rule set, a "T" always follows an "M" or a "P" must always follow an "S." For each letter group, a participant would have to guess whether that group was a member of family A or family B—that is, whether it had been generated by rule set A or rule set B.

After a while, Reber's participants got better and better at this. By itself, this isn't remarkable. People tend to improve at most things if they keep working at them. What was surprising was that they weren't able to explain the rules that described each of these families. They knew that they had somehow learned the rules. They just couldn't bring them to awareness. Furthermore, when they were instructed to consciously try to figure out the rules while making their judgments, they did worse than when they used a more passive and intuitive guessing strategy. A deliberate, analytical mind-set squashed their intuition.

This remarkable ability to intuitively learn patterns is more than just a way for you to acquire information about your environment. It can help you to regulate your thoughts and plan your future creative efforts.

THE CREATIVE GUIDE

Perhaps you've been working on a problem for a while and you're stumped. There's a potential cost to continuing. For example, if you're a scientist, you might spend both time and scarce research funds trying to solve a problem that's ultimately unsolvable. How can you know whether you should keep plugging away?

Consider again the case of Judah Folkman. As a young navy surgeon trying to develop a blood substitute with a long shelf life, he observed that blood vessels start growing toward tumor cells. Both before and after this finding, others had observed similar phenomena and simply ignored them or shrugged them off. But Folkman immediately knew that this was important, *very* important. He was possessed by this seminal observation to the point that it dominated his research for more than four decades. Over the years, he had a series of insights that led him to understand that he had discovered what he would later call "angiogenesis," the process by which new blood vessels grow, in this case toward a tumor to feed it. Folkman's vision of angiogenesis as a key step in the progression of cancer and other diseases—a step that could potentially be disrupted—was for many years ignored or derided by the research community and by agencies with the resources to fund his research. In spite of the ridicule and lack of funding, he persisted. Every time he reached a dead end, he continued to hack away at the problem, leading some colleagues to believe that he was a mediocre scientist because he was wasting time on an idea that would never pan out. But Folkman pressed on because he was convinced of the validity of his vision. He didn't know exactly how he would achieve it, but he believed—he *felt*—that there was a path, even though he was periodically blocked by apparently insurmountable technical or theoretical problems. He persisted because he always sensed that the next insight was just around the corner.

Judah Folkman's story shows how intuition can play an important role in regulating your creative efforts. Even if you don't know what your next insight will be, your brain might. If the idea is already brewing, your brain may tell you, through the experience of intuition, to hang on because it's preparing to send you an idea that will arrive in due course.

However, intuition can also mess with you. What if a hunch is simply wrong? Like every other mental ability, it may be in error or reflect wishful thinking rather than heightened self-awareness. When this happens, you might engage in an endless quest for an idea that doesn't exist, such as the fountain of youth.

A wrong hunch isn't the only pitfall. The absence of an intuition can be just as wasteful as a false one. Perhaps you avoided a problem because you didn't feel that there was any hope of finding a solution in a reasonable amount of time. So you rejected the challenge. Then later, perhaps too late, the solution came to you as a sudden insight.

Intuition can be a great asset but carries risks. Let's examine how and when it works.

FLYING BLIND

The idea that insight is a unique process that is different from analytical thought used to be fighting words among cognitive psychologists. Some researchers argued that what we call "insight" is just normal analytic thought, albeit with an emotional burst adorning the solution.

Janet Metcalfe entered this debate with a trailblazing series of studies. She proposed that if insight really were sudden, then before you have the insight, you shouldn't have any sense that you're close to achieving the solution. According to this logic, a true insight can't be presaged by an intuition because you haven't yet acquired the new perspective that would enable you to sense that the solution is near. Only after the insight have you attained the bird's-eye view—the "Archimedean point"—for surveying the vista.

In contrast, analytic solving taps well-known procedures and your current understanding of the problem. As you work on a problem in a deliberate, step-by-step fashion and pass each milestone, you can tell that you're getting closer and closer to the solution. Here's an analogy: You can't know how long it will take you to cook dinner until you have the recipe for the dish. But once you have the recipe, you can make an educated guess about how long it will take to prepare. And while you're cooking it, you can refine your estimate.

Another analogy is the children's game in which one child hides something and another one has to find it. While searching for the hidden object, the child may ask her friend whether she's getting "warmer" or "colder." If you already know that the object is in the second drawer of the dresser in the bedroom, then it's obvious whether you're warm or cold, because you know how far you are from the dresser. In analytic problem solving, you know whether you're getting warmer or colder because you already know the steps to take to get to the solution. However, if insightful solving does involve unconscious processing followed by a sudden realization of the solution, then asking if you're getting warmer should be meaningless. It would be like asking your playmate how warm you are when you're in different rooms and neither of you knows where the object is.

Metcalfe gave her participants both insight and analytic problems to solve. While they were working on these problems, she asked them every fifteen seconds how warm they thought they were—that is, how close they thought they were to the solution. For analytic problems, the participants started out cold and gradually felt warmer and warmer until they reached the solution. For insight problems, they didn't feel warm until just before they actually solved them. But when they felt gradually increasing warmth preceding an insight solution, their answers were usually wrong.

Metcalfe's findings make perfect sense. You can't see what's around the corner until you've turned it. If you have to reinterpret a problem to solve it, then you won't be able to judge how close you are to the solution until after you've reinterpreted it. In contrast, analytic thought travels a straight, familiar, well-lit road studded with signs and milestones that tell you how close you are to your destination.

But this view misses something important. Recall that our research revealed the existence of a "brain blink" that reduces visual inputs just before an insight but not before an analytic solution. For this to happen, the brain must somehow already "know" that an unconscious solution is ready to burst into awareness as an aha moment. Otherwise, how would the brain know to briefly blink the mind's eye to help you find the solution? There has to be some kind of knowledge about the insight before it arrives. But what kind of knowledge?

INKLING

Researchers used a version of the remote associates problems to tackle this question. Only half of the problems had a solution. For each puzzle, a participant's job was to quickly judge whether it's solvable at all—its "semantic coherence"—rather than what the solution is. Each person had only 1.5 seconds to make this judgment. This is too short to solve one of these problems. Under these circumstances, all one can do is to make a quick, intuitive guess about whether a solution exists.

Canadian and German scientific teams showed that people have better-than-chance intuitions about the solvability of remote associates problems. And this ability isn't confined to verbal puzzles. Participants also had better-than-chance intuitions about which jigsaw puzzles could be rearranged to show pictures of real objects. These findings mean that, at least some of the time, the solutions were unconsciously activated in people's minds and their presence could be sensed.

In contrast, Metcalfe's U.S. studies of intuition showed no evidence of intuition about an impending insight. In her studies, people's insights seem to have come out of the blue. How can we reconcile these contradictory findings without jumping to the conclusion that Canadians and Germans are intuitive, but Americans aren't? There is a more plausible explanation—one that reveals a crucial property of intuition. This explanation is an important part of knowing which intuitions to take seriously and which to take with a grain of salt.

REMOTE SENSING

Think of the movie *Jaws*. The shark is down there, somewhere, but you don't know exactly where, because the ocean is murky and you can't see more than a couple of feet deep. At some point, it will suddenly attack. You just don't know when. The shark will be visible for only a moment—just before it strikes.

In Metcalfe's U.S. experiments, people already knew that each problem had a solution. Their job was to judge how close it was to breaking through to awareness, which they couldn't do until the insight was imminent. The German and Canadian participants faced a different challenge: judging whether a solution even existed. They were able to do this with reasonable accuracy. So these two groups of studies don't contradict each other; they were tapping different aspects of intuition: sensing *presence* and sensing *closeness*. Even if you can sense that the solution is buried in your unconscious mind, this doesn't mean that you can tell when it might surface.

Outside the laboratory, the difference between these two forms of intuition can lead to frustrating decisions. You may feel that the solution is down there, somewhere, but you can't tell if it will materialize in a minute, a day, a month, a year—or *never*. Should you keep trying? The answer to this question depends on both the value of the solution and the odds of success. Spending a decade, or a whole career, working on a cure for Alzheimer's disease is certainly worthwhile, even if the chance of success is low, because the effort is noble and the potential cure so important. But not every problem may merit such an investment of time without a likely payoff. It's a judgment call.

MOODY AND FLUENT

So intuition is real, but limited in what it can tell you. And even if you can sense that a solution exists, this feeling can't always be trusted because intuition can be unstable. Fortunately, we're beginning to clarify intuition's strengths and weaknesses, which allows us to gain a better understanding of when to trust it. There are two key factors to consider.

Your mood is one of them. Cognitive psychologist Annette Bolte and her team put people in a positive or negative mood by asking them to remember happy or sad events. Then she had them quickly guess which remote associates problems had solutions. Their intuitions were sharpened by a good mood and dulled by a bad one. So, get up on the wrong side of the bed, and you could be clueless all day.

The other key factor is "fluency," which is the ease of a mental process. For example, you can increase the fluency of your reading simply by using a clearer type font or by increasing the visual contrast of the words on a screen. Repetition or practice also increases fluency: The second time you look at a complex picture, you can perceive it more quickly and easily than the first time. But turning up the lights and practicing aren't the only ways to increase your mental fluency.

Cognitive psychologists Sascha Topolinski and Fritz Strack found that when you look at a remote associates problem such as "pine"/"crab"/"sauce," you can actually read the three words more quickly when a single solution word ("apple") exists. When you read the triplet, these words quickly and unconsciously prime the solution word, which, in turn, quickly primes the problem words (see figure 10.3). This makes the triplet easier and faster to read. It's like preheating an oven so it won't take you as much time to cook your food.

None of this occurs for triplets like "gravy"/"elm"/"lobster," because they don't have a single unconscious solution word that connects with the problem words. That's why people read such triplets more slowly.

More generally, when you see a problem whose parts are connected to an unconscious solution, then the problem is easier to perceive. The problem seems more natural, pleasing, and right. But fluency is just one part of the intuition apparatus. It can make it easier to take in problems that have solutions, but it doesn't directly lead to the sense that a problem has a solution. Fluency's contribution to intuition is indirect. It's just the first step. The next involves emotion. This is the tricky part.

THE FACE OF INTUITION: KEEP SMILING

People *like it* when a problem—when *anything*—is easy to think about. Things that you process fluently trigger in you a subtle, tem-

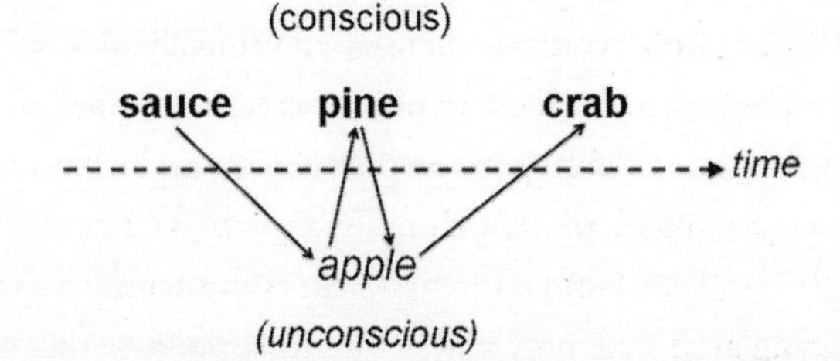

FIGURE 10.3: How a common associate ("apple") primes a word triplet, making the triplet easier to read. A person reads the words "sauce," "pine," and "crab" sequentially. Reading "sauce" first primes the unconscious solution word "apple" (among other words), which in turn primes "pine" and "crab," making those words easier and faster to read. Reading "pine" re-primes "apple," which contributes further to the priming of "crab."

porary burst of joy. Your brain interprets this pleasant feeling as an intuition that the parts of a problem have underlying connections to one another and to a solution. Topolinski and Strack demonstrated this in several ways, though their most audacious results came from studying people's facial expressions.

Your facial muscles not only express your emotions; they also influence them. When you feel happy, then you smile; when you smile, then you will feel a bit happier. The researchers used this principle in an ingenious experiment. They placed sensors on people's faces to monitor subtle tension in particular muscles. (They deceived their participants with a cover story to hide the true purpose of the sensors.) When people saw solvable word triplets, there was a slight tightening of a key smile muscle; when they saw unsolvable triplets, frown muscles tensed. These findings show that problems that have solutions give you a fleeting boost; those without solutions give you a temporary downer. Topolinski and Strack could actually read these intuitions on their participants' faces. But they didn't stop there.

Try this: Open your mouth a little and then hold a pen in place horizontally with your lips for a few seconds without clenching your teeth. How do you feel?

That exercise forces you to tighten the same muscles that you use to smile, causing a tiny, easy-to-miss wave of positive emotion. The researchers used that technique to induce a subtle good mood in their participants (who were tricked by a cover story). At other times, they attached golf tees to the inner ends of people's eyebrows and asked them to make the two tees touch each other by bringing their eyebrows together. That facial expression, akin to a scowl, induces negative emotion. The artificial smiles lured people into thinking that word triplets had solutions; artificial frowns led people to think that they didn't.

Thought and emotion are blended. A positive mood energizes a person's unconscious associations so that when she views a coherent problem, the problem's parts prime the solution and one another more strongly, making them easier to perceive. This triggers a little burst of pleasure that she interprets as an intuition that a solution is present. It also encourages her to continue working on the problem until the solution proclaims itself in the form of an insight.

When Dr. Susan Hasselquist was confronted with Rebecca Woodings's collection of symptoms, some of these symptoms were unconsciously associated in the doctor's mind with the correct diagnosis of Addison's disease. In turn, this diagnosis primed her perceptions of the patient, highlighting the key symptoms and downplaying the irrelevant coughing. Thus, as she examined Woodings, her perceptions of the patient and the most important symptoms were fluent, evoking a somewhat pleasurable feeling that the doctor interpreted as an intuition of impending diagnostic insight. This encouraged her to persist until the insight emerged.

This understanding of intuition does a good job of explaining how we can sense the presence of a solution or an idea. However, as a theory, it's incomplete because it doesn't deal with other kinds of intuitions.

One morning in 2001, Jackie Larsen of Grand Marais, Minnesota, went to a meeting of her prayer group. Upon leaving, she met Christopher Bono outside. Bono seemed like a nice young man, well groomed and polite. He told her that he was on his way to meet friends in Thunder Bay, but his car had broken down and now he was looking for a ride to get there. Larsen told him to come along to her nearby shop, where they could find his friends' phone numbers in the phone book and call them to pick him up.

Larsen was undoubtedly in a good mood after her prayer meeting. Nevertheless, she experienced a sharp pain in her stomach. She didn't know what was wrong—Bono seemed respectable enough—but she knew that *something* was wrong. She decided to keep him talking outside on the street so that she didn't have to let him into her shop. At one point, she blurted, "I am a mother, and I have to talk to you like a mother." Bono stared at her and said, "I don't know where my mother is." Larsen then deflected Bono and sent him to talk to her pastor. While he was doing this, she called the police to tell them about the young man who gave her a bad feeling. They traced his license plates and found that the car was registered in Illinois—to his mother. They called the police in Illinois and asked them to check on Bono's mother. When the police knocked on her door, no one answered. Upon entering, they saw blood all over the apartment. Lucia Bono was in the bathtub—dead.

Jackie Larsen showed acute intuition on that day. Even though she didn't know specifically what was wrong or why she felt that way, she felt confident enough that something was wrong that she felt impelled to act. Her intuition enabled the police to discover a murder and arrest Bono. It may also have saved her life.

A good mood would have energized Larsen's mental associations and readied her for an intuition. When she met Christopher Bono, she had a hunch, but she experienced it as a burst of negative rather than positive emotion. Her gut told her that something was wrong. Perhaps this was triggered by a subtle incongruency in Bono's appearance or demeanor—nothing she could put her finger on, but enough that it caused a decided *lack* of fluency in her perceptions of Bono. Things didn't seem to fit, and this made her feel bad.

Larsen's intuition was different from those experienced by participants in the laboratory tasks that we described. Her hunch was based on a surge of negative rather than positive emotion. This suggests that intuitions can be caused either by bursts of positive emotion from fluency or by bursts of negative emotion from disfluency. Either way, emotional surges, whether positive or negative, seem to have their strongest influence on intuition when they occur against the backdrop of an ongoing positive mood because that's when a person's associations are most energized.

When you encounter a problem that has a solution, it can look different. It can literally *feel* different. But throw a monkey wrench into the brain's emotional system, and intuition gets scrambled in the process.

THE HEAVY HAND OF THOUGHT

Because intuition is a delicate, complicated process, it can go wrong in many ways. For example, intuition is so fragile that analytic thought can completely crush it. Topolinski and Strack's participants showed better intuition when they passively looked at the word triplets and simply guessed than when they deliberately tried to figure out the solutions. The loud shouts of analytic thought overwhelm the subtle whispers of intuition.

The researchers also discovered a particularly insidious method for tripping up intuition—by manipulating people's understanding of their own emotions. They led people to a noisy room next to a cafeteria and gave them headphones to listen to bland music to block out the noise. (The cafeteria noise was the cover story that justified the headphones and music.) The participants then guessed which word triplets had solutions. The ones who were told that the bland music would influence their emotions assumed that their emotional changes were caused by the music and not by the word triplets. This robbed them of their intuitive ability to judge which problems had solutions.

Thus, a burst of emotion can inform you about the presence of an unconscious solution, but only if you correctly understand why you are experiencing the emotion. If Jackie Larsen had eaten some questionable clams on the day she met Christopher Bono, she might not have taken the sharp pain in her gut seriously.

TRUSTING YOUR GUT

An ongoing negative mood signals that something is wrong. Danger lurks. This invokes an analytic mind-set and squashes your remote associations. You consciously ruminate over every detail and deliberately reason through everything because, when a situation is threatening, errors can't be tolerated. But when you feel that your environment is safe, your mood is good. This grants you permission to engage in a more intuitive and creative style of thought. This freewheeling cognitive style is riskier. It's "iffy." We've seen how fragile intuition is. Impressions can be vague. Jumping to conclusions can lead to errors. But an error here or there is tolerable when there is no serious threat. If you feel safe, why not take a chance and throw caution to the winds? Who knows what interesting things might happen?

Many years ago, during the summer vacation before John's first year of college, he had a part-time summer job in a department store that was a twenty-minute drive away from his home. One morning, he followed his usual routine of eating breakfast, showering, dressing, and so forth. He got in the car and started to drive to work. He was in a good mood because he had graduated from high school and was thinking about the adventure of starting college. However, after driving for about ten minutes, John was seized by the feeling that something was amiss. *He had forgotten something.* But what did he forget? He quickly ran through a mental checklist. His wallet was in his pocket. The gas tank was full. Nothing seemed to be missing. Even though he wasn't able to put his finger on what could be wrong, the feeling just wouldn't go away. Because he couldn't think of what he might have forgotten, his analytical half told him that this feeling was vaporous nonsense.

However, John's intuitive half just had to know. It demanded that he go back home and check, even if this made him late for work. So he took the chance. It was an experiment—he had to know whether his trust in this feeling was well placed.

He arrived back home and went to the door. He didn't want to bother his mother, so instead of ringing the doorbell he reached for his key chain to open the door. He suddenly realized that the key chain had his car key but not his house key. Then he remembered that he had removed the house key from his key chain a few days earlier to give the key chain with the car key to a repair shop. After picking up the car from the shop, he had neglected to put his house key back on the key chain and had left it inside the house when he went to work that morning. If he hadn't returned and realized that he had forgotten his house key, then he would have found himself locked out of the house when he came home after work because no one would have been home to let him in. Some part of him knew that he had forgotten something, even though he couldn't consciously figure out what it was. But he had been in a good mood, so he took the chance and trusted that his brain knew what it was doing. His good mood not only made the intuition possible, but it also inclined him to trust it. This kind of trust is another key ingredient to the effective use of intuition.

THE MEANING OF LIFE

What did the disaster of Hurricane Katrina mean to you? How about the massive oil spill from British Petroleum's offshore well in the Gulf of Mexico? The tsunami and nuclear reactor disaster in Japan? Is there a deeper significance to these events, or were they random and senseless?

Consider the following sayings. How meaningful are they to you?

> "Sometimes it's necessary to go a long distance out of the way in order to come back a short distance correctly."

> "Think like a man of action; act like a man of thought."

> "No matter where you go or what you do, you live your entire life within the confines of your head."

Now consider the following Zen koans. What, if anything, do they mean to you?

> "If you understand, things are just as they are. . . . If you do not understand, things are just as they are."

> "One day as Manjusri stood outside the gate, the Buddha called to him, 'Manjusri, Manjusri, why do you not enter?' Manjusri replied, 'I do not see myself as outside. Why enter?' "

> "Two monks were arguing about the temple flag waving in the wind. One said, 'The flag moves.' The other said, 'The wind moves.' They argued back and forth but could not agree. Huineng, the sixth patriarch, said: 'Gentlemen! It is not the flag that moves. It is not the wind that moves. It is your mind that moves.' "

If you question people, you'll find that some see meaning everywhere, in events like the Japanese tsunami and in cryptic sayings like those above. They will give you impassioned explanations of the significance of such things. Other people deny any inherent meaning. "Stuff just happens. Live with it."

What accounts for such differences in attitude?

Social psychologist Joshua Hicks and his collaborators tackled this question. They presented such sayings and abstract drawings to people and asked them to rate how meaningful the participants felt that they were. They also asked participants to rate the meaningfulness of particular life events such as Hurricane Katrina. Importantly, they also asked their participants to fill out questionnaires that assessed their mood and aspects of their personalities.

By now, it shouldn't be surprising that they found that people tend to sense meaning in events when they are in a positive mood and see things as random and meaningless when in a negative mood. But there is another key factor that regulates the relationship between mood and intuition.

The questionnaire that participants had filled out assessed a psychological characteristic called "faith in intuition," which is the extent to which a person has intuitions and trusts them. This questionnaire consists of a series of statements with which a person indicates agreement or disagreement. For each statement, she rates the strength of her endorsement or rejection. Example items include:

"I believe in trusting my hunches."

"I tend to use my heart as a guide for my actions."

"I rely on my intuitive impressions."

"I trust my initial feelings about people."

Hicks found that a positive mood contributed to the sense that events have meaning—but only for those people who scored high in faith in intuition. For those with little or no faith in intuition, a positive mood didn't enhance their sense of meaning in the world. So mood influences intuition, but not equally for everyone.

One potential criticism of this finding is that when people are in a good mood and trust their intuitions, they are essentially putting on rose-colored glasses. They may be biased to think that the world has meaning, even when it doesn't. No one can objectively prove that a tsunami or a Zen koan has any real meaning, so it's impossible to prove that the tendency to see meaning in such things isn't just an expression of subjective bias rather than objective perceptiveness. If this is just bias, then we shouldn't believe that intuitions have any validity or significance. They are just as meaningless as the events that evoked them.

The researchers addressed this point with a laboratory task that does have objective solutions: the semantic coherence task. The results were the same. People who trusted their intuition *and* were in a good mood did better at quickly judging which word triplets had solutions and which didn't. So intuition can be enhanced by a positive mood. But this doesn't work for everyone. It's not enough to have accurate intuitions. You also have to trust them.

And like mood, this trust is a malleable thing.

It's been shown that Intuitives tend toward belief in God. Intuitives often sense a connectedness and meaning in the world that they attribute to God's influence. This relationship between intuitiveness and religious belief is independent of IQ, personality, and other factors. But that doesn't mean that it's fixed in all Intuitives. When researchers asked participants of a study to remember instances from their past in which intuition had served them well, this actually *increased* their belief in God. Conversely, recalling successful examples of critical analytic thought decreased it. So not only might some people's faith be temporarily bolstered or shaken by recalling events from their past, but also their faith in intuition might be strengthened or sapped, with effects that ripple through their understanding of the universe and themselves.

For some, it's natural to simply "trust one's gut." For others, it's difficult to yield to their intuitions. The question is why. One possibility is that faith in intuition is simply a matter of faith. A person may have no objective evidence that his intuition is reliable but may trust it anyway because it's part of his belief system or self-image.

However, we all know people who have consistently bad intuitions but nevertheless continue to trust them, sometimes with disastrous consequences. Indeed, some political and business leaders can be counted among this group. Other people who put too much faith in their poor intuitive ability follow their gut because their analytical abilities are even worse. They choose intuition as the lesser of two evils when the smarter choice (which they don't have the sense to make) would be to let someone else do the thinking for them.

But this doesn't explain all people's choices.

A DIFFERENT KIND OF SMARTS

Until recently, it was assumed that differences in cognitive ability were almost entirely due to analytic thought—the kind of thought measured by IQ tests. The idea was that people differ greatly in analytic ability; they don't differ much in the kind of unconscious associative processes on which intuition is based. We now know that this view is wrong.

Psychologist Scott Barry Kaufman and his colleagues gave their participants a battery of cognitive tests, including both tests of analytical ability and tests of unconscious intuitive thought. Just as there are substantial differences among people in analytic intelligence, they also found substantial differences in intuitive ability. And these two types of cognitive ability weren't strongly related to each other—just because you're good at one doesn't mean that you're good at the other.

Intuitives—people who are high in intuitive ability and are close relatives of our Insightfuls—have a particular type of personality. They are impulsive thinkers who appear to shoot from the hip. This is because the intuitions that they rely on can often be quicker than the lumbering methodical thought of Analysts. Intuitives are also high in a personality trait known as "openness," which means that they tend to focus on their experiences of patterns and feelings rather than on their analytic thoughts.

The fact that some people really are more intuitive than others goes a long way toward explaining why some trust their intuitions while others rely more on analytic thought. Most people generally have a sense of what they are good at. Not everyone has accurate self-knowledge, of course. People tend to be overconfident about their cognitive abilities, which is why students frequently wrap up a test feeling that they aced it when they, well, didn't. But most people do seem to have some appreciation for their intellectual strengths and weaknesses. They go with what seems to work best for them, which, for Intuitives, are the greased associative pathways that allow ideas to ping one another in an unconscious chain reaction.

But if this pinging ends there and never culminates in a conscious aha moment, then all that the Intuitive is left with are vague feelings—whispers from the basement of the mind. We've all met people like this. They have good instincts but don't understand and can't explain why they act and feel the way they do. Intuitives of this type haven't yet gotten in touch with their inner Insightful. Their feelings of impending enlightenment don't culminate in an aha moment.

Finally, even if you are an Intuitive, a cautionary note is in order. Intuition can be a powerful way to know the world. Trusting your hunches often works. It can let you sense the impressions, patterns, and ideas percolating in your unconscious mind. But many factors can create a false confidence that an intuition is valid. So when you have an intuition, don't make an impulsive decision to either follow or reject it. Ask yourself *why* you are having that hunch, at that time, and at that place. The answers to these questions will help you to decide whether to trust a particular intuition. Remember, your brain knows more than you do, but it doesn't know everything.

11

THE INSIGHTFUL AND THE ANALYST

Whoever cannot seek the unforeseen sees nothing, for the known way is an impasse.

—Heraclitus, *Fragments*

In a commencement address at Oberlin College in 2002, Judah Folkman recounted an experience that he had had in high school. His chemistry teacher had assigned him the task of identifying several substances, using laboratory methods that the students had learned in class. Folkman correctly identified each of these substances until the last one, a clear fluid, stumped him. None of the tests seemed to work. Sensing a trick, he declared that it must be plain water. The teacher told him that it wasn't good enough to guess that it was water—he had to present some kind of supporting evidence. Folkman thought for a moment and then asked his chemistry teacher to unlock the physics lab. Using wires, a lightbulb, and a battery from the physics lab, Folkman showed that the liquid did not conduct electricity until salt was added, as would be expected with salt water.

This satisfied his teacher, who gave him extra points for this idea. Folkman described what happened next: "I started to walk home. Just before crossing the street, I was struck by a sudden thought, and almost by a car. *My gosh, chemistry is one subject and physics is another. They are taught by different teachers on different floors at different times from different textbooks, but they are connected. . . . They are not separate in nature, only in school. So, aha! Nature is not arranged in the same way that schools are.* Later I learned, much later, that this experience is called an 'aha moment' by scientists, and by philosophers, an epiphany."

Even as a child, Folkman saw similarities and connections where others saw only boundaries and barriers. This was his nature.

During his time in the U.S. Navy, Folkman heard that the crew of a small submarine had died when the sub became tangled in some cables and lost power. Naturally curious, he learned that the artificial atmosphere on the submarine had contained a substantial amount of helium. Knowing that helium conducts heat very efficiently, he inferred that the crew members had died from hypothermia when the sub lost power and the helium drained the heat from their bodies. An idea clicked in his mind. Aided by his knack for building things, he invented a special blanket that used helium to cool a patient during surgery.

Judah Folkman was an archetypal Insightful. His ability to see connections and analogies powered the insights on which his long research career was based. But not all scientists, not even very productive ones, are granted such insights. Folkman recognized this, and once said, "Every once in a while . . . you realize that you know something . . . that no one in history has ever known and you can understand something about how nature works. That doesn't happen to most . . . scientists, and if it does, it's a blessing, and if it happens more than twice, it's a miracle. And when it happens, it's a very big high."

Fred Rosen, Folkman's colleague at Harvard Medical School,

said, "Science is an extraordinarily disciplined business, and occasionally people come along like Judah who break the rules, and people's rigidities. It happens with people who are very, very creative, people who have vivid imaginations."

Who are these "very, very creative people who have vivid imaginations"? For that matter, why do people differ at all in their styles of thought and ways of approaching a problem?

DIFFERENT STROKES

You probably know some Insightfuls: people who seem to thrash around, haphazardly generating ideas and apparently going off on tangents until they eventually solve a problem with a flash of insight that allows them to make a startling connection. You probably also know some Analysts, people with a keen ability to analyze a problem and systematically work out a step-by-step plan for deriving a solution. Less-organized Insightfuls often admire this ability because of its thoroughness, clarity, and apparent objectivity; Analysts—at least the ones with some sense—often admire Insightfuls' ability to see the forest instead of the trees and to see this forest from unusual angles.

Most people can, and to some extent do, use both of these approaches. A pure type probably doesn't exist; each person falls somewhere on an analytic-insightful continuum. Yet many—perhaps most—people tend to gravitate toward one of these styles, finding their particular approach to be more comfortable or natural. These individuals offer important clues for understanding and enhancing insightful and analytic thought, so let's take a look at how their brains work.

We recorded people's EEGs while they each sat and relaxed for a few minutes. They were given no task and didn't know what they would be doing next. They just let their thoughts flow. Cognitive neuroscientists call this the "resting state." Afterward, they were given a series of anagrams to solve. Unsurprisingly, some participants tended to solve more anagrams analytically, while others solved more with insight. We compared the resting-state EEGs of these Analysts and Insightfuls to discover whether their brains differed in important ways even when they weren't working on a problem. Now, rather than examining differences between analytic and insightful thinking, we were investigating differences between analytic and insightful *thinkers*.

We expected these groups to differ, but we were astonished at the magnitude and scope of what we found. To understand these differences, we organized our findings around three principles.

Taking It All In

First, creative people, including those who do well on paper-and-pencil tests of creativity as well as those with a track record of real-world creative achievement, tend to have habitually unfocused attention. This can make it difficult for them to shut out distractions, such as the sound of a nearby conversation, when they are trying to work. It's not that they can't focus when they need to. In fact, for relatively short periods, they can focus at least as well as other people, perhaps better. But this isn't their natural state, so it's a bit harder for them to sustain it.

Our brain wave findings illustrate this principle. Figure 11.1 shows a map of EEG brain wave activity measured at the back of the head. It shows a major difference between our Insightfuls and Analysts. These electrodes (shown as dark dots) lie over the visual cortex, which is in the back of the brain. As shown by the white oval on the map, we detected more visual cortex activity for Insightfuls compared with the Analysts. Even in a resting state, the Insightfuls' brains were doing more visual information processing.

On the surface, this would seem to contradict some of our other research findings. Recall that just before solving a problem by insight, people's brains briefly shut out information from the environment.

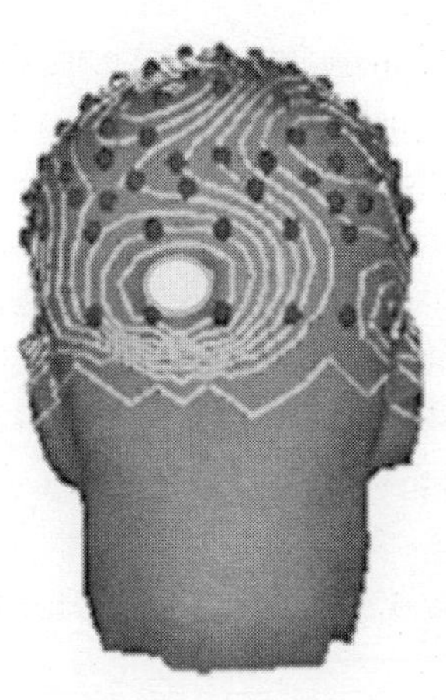

FIGURE 11.1: EEG map of the back of the head shows greater neural activity (marked by the white oval) over the visual cortex for Insightfuls than for Analysts while at rest. *John Kounios and Mark Beeman*

Reducing distractions from one's surroundings helps you focus attention inwardly, find the weakly activated idea, and bring it to awareness. That occurs while a person is actively working on a problem and also while anticipating a problem. But it doesn't happen during rest. While an Insightful is in a resting state, which includes periods of incubation, the perceptual floodgates are open and everything rushes in. Insightfuls' relatively scattered and far-ranging minds tend to take in lots of seemingly irrelevant pieces of information, any one of which could trigger an insight. Another way to think of this is that Insightfuls are more inwardly focused while working on a problem and more outwardly (and broadly) focused when at rest.

The eyes of the Insightful aren't the only things that are wide open. So, too, is his mind. Remember that perceptual attention is closely linked to conceptual attention. Factors that broaden your attention to your surroundings, such as a positive mood, have the same effect on the scope of your thinking. Besides taking in lots of seemingly unrelated things, the diffuse mind also entertains seemingly unrelated ideas. An Insightful might be sitting at the dining room table, eating dinner, and notice the table, which incites the remote association "water table," which reminds him to water the lawn, which makes him think of golf, which makes him think of the dimples on golf balls, which reminds him of dimples on faces and how so-and-so lost face at the office meeting today, which makes him think of a more effective way of handling an office issue, and so forth. All the while, the more focused Analyst would be thinking about whether to finish the broccoli on the plate in front of him.

Doing the Right Thing

We also predicted, and found, that while in the resting state, the Insightfuls had more right-hemisphere activity and less left-hemisphere activity than the Analysts. Our participants were not performing a task and didn't know that they were about to tackle a series of problems, so this brain activity reflected their own spontaneous thought rather than any directions from the experimenter. This right-hemisphere activity suggests that Insightfuls' free thought involves looser associations than those of the Analysts.

Going Rogue

The third principle is "cognitive control." Tight cognitive control imposes close supervision on thought, keeping us on message and focused on the dominant possibilities in any situation. This is the way of the Analyst.

The frontal lobe is the major source of cognitive control in the human brain. It exerts "top-down" control to organize our thoughts, emotions, and behavior by highlighting what's deemed important and by dampening what's irrelevant or undesirable. In contrast, "bottom-up" processing is a grassroots movement. It occurs when something in the environment grabs control of the brain, as when you reflexively attend to a loud, unexpected sound, such as a door slamming. Bottom-up processing is dominated by what's happening in the world around you; top-down control is an expression of your goals, plans, and values.

We examined top-down control by looking at communication between the frontal lobe and other parts of the brain. Our Analysts showed stronger communication between the front and back of the brain than did our Insightfuls, signifying the Analysts' greater top-down control (see figure 11.2).

Intervention by the frontal lobe is usually necessary to restrict the vast range of possibilities that a person would otherwise have to consider. But recall Carlo Reverberi's study of patients with frontal lobe damage that we described in Chapter 3. They were actually better

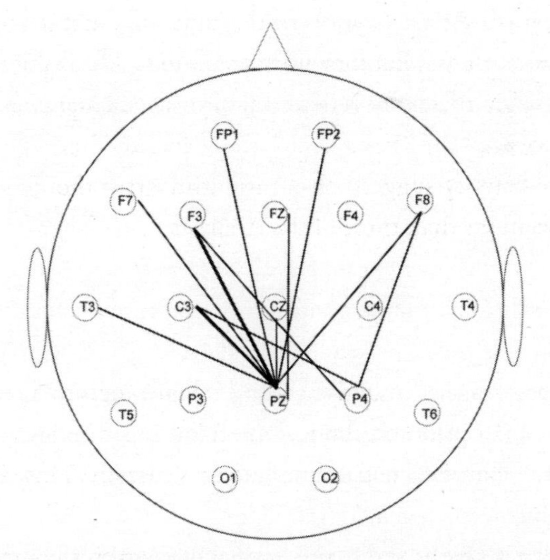

FIGURE 11.2: EEG alpha "coherence" map showing greater front-to-back communication between brain areas of Analysts compared with Insightfuls. The map represents a top view of the head with the nose at the top and the ears on the sides. The small circles represent the locations of the EEG electrodes. The straight lines connecting some electrodes show pairs of electrodes with greater synchronous brain wave activity for Analysts. All of these lines show communication between the front and the back of the brain, and all of this communication was stronger in the Analysts than in the Insightfuls. *John Kounios and Mark Beeman*

than people with intact brains at solving problems that usually require insight. Frontal lobe damage removed these patients' cognitive blinders, allowing them to consider unusual possibilities. Our Insightfuls didn't have such damage, but their frontal lobes were apparently less bossy than those of the Analysts. So, even when not engaged in a task, an Analyst's frontal lobe calls the shots. The Insightful's brain is more disorderly and rebellious.

Resting-state brain waves tend to be fairly stable over time and are strongly influenced by one's genetic makeup and brain structure. This suggests that Insightfuls and Analysts should be fairly consistent in their cognitive styles. A recent study in John's lab supports this idea—EEG patterns predict insightful solving, even when the EEGs are recorded several days before participants tackle a set of problems.

But there is another source of evidence for the idea that insightfulness is a relatively stable characteristic. It seems that insightfulness is a blood relative of madness, and madness is firmly rooted in one's biology.

MADDENING CREATIVITY

It would be difficult to find an informed person who would disagree with the notion that Sir Isaac Newton (1643–1727) was one of the greatest geniuses ever to walk among mere mortals. His contributions to physics (optics and classical mechanics) and mathematics (calculus) are monumental and seminal, securing him a permanent place in the pantheon of great achievements.

Sir Isaac also had the distinction of experiencing the second most celebrated aha moment in scientific history (after Archimedes, a genius of comparable stature). As the famous story goes, in 1666, to avoid a plague that was ravaging the cities, Newton left Cambridge to return to his childhood country home in Lincolnshire. While sitting (or meandering, according to one version) in his garden, lost in thought, he saw an apple fall from a tree. (Existing accounts don't mention the apple hitting him in the head, an embellishment added by later generations.) Noting that the apple fell perpendicularly to the ground, he made the mental leap that objects must exert a pulling force on each other, and that the strength of the force is proportional to the mass of the object. So the earth pulls the apple toward its center, and the apple pulls the earth toward its center. Of course, the earth has much greater mass, so it's the apple that moves, not the earth. Furthermore, the same mysterious force must be responsible for keeping the moon in its orbit around the earth so that it doesn't fly off into space. This force is gravity.

We don't know with certainty that this particular anecdote is true. Perhaps it was just a colorful story that Sir Isaac fabricated to entertain people (though he was not known for being a particularly entertaining person). But whatever the facticity of this anecdote, aha moments were clearly not unusual for Sir Isaac. His secretary wrote, "When he has sometimes taken a Turn or two [in the garden], has made a sudden stand, turn'd himself about, run up ye Stairs, like another Alchimedes [*sic*], with an ευϱηκα [eureka], fall to write on his Desk standing, without giving himself the Leasure to draw a Chair to sit down in." As author James Gleick once said, "He was somebody who seemed to pull ideas out of nowhere." Sir Isaac obviously had advanced analytical abilities. Nevertheless, at his core, he was an Insightful.

He was also, to put it mildly, an unusual person. As physician-historian Milo Keynes notes, "He was a suspicious and . . . paranoid man. . . . He had a sensitive, prickly personality at the edge of normality, which could tip over into psychosis, if the conditions were right and sufficiently emotionally upsetting: that is, he occasionally lost the capacity for reality testing." He had at least one "nervous breakdown."

We will never know the exact nature of Newton's occasional lapses into mental illness. But his case was not exceptional. Throughout history, observers have noted a connection between creative genius and insanity. Another example is the celebrated mathematician and eventual Nobel laureate John Nash. At the height of his mathematical powers, Nash's strangeness descended into schizophrenia that lasted for many years before its worst symptoms receded.

What is the connection between creative insight and madness? The answer to this question sheds light on the stability of insightfulness. There is reason to think that creativity and mental illness are related and that they are both expressions of something that is relatively stable—one's genes.

JUST A LITTLE OFF

Schizophrenia is a devastating psychiatric disease characterized by disordered thought and very loose mental associations. Though environmental factors regulate its expression, it is inherited and conferred by many genes rather than a single one.

Schizophrenics are seriously ill, and their symptoms can interfere with their social interactions, so they tend to have fewer children than healthy nonschizophrenics. This raises a difficult question. In theory, reduced procreation should cause schizophrenia-related genes to disappear with the passing of generations. Why, then, haven't millennia of Darwinian natural selection simply bred schizophrenia genes out of existence? An answer to this question can be found in an analogy between schizophrenia and another genetic disease.

Sickle-cell anemia causes a malformation of red blood cells and an alteration of their oxygen-carrying hemoglobin molecules. This interferes with red blood cells' ability to transport oxygen through the bloodstream. Sickle-cell anemia causes much suffering and can greatly shorten a patient's life span.

A person must inherit two sickle-cell genes to get the disease, one provided by each parent. If a person inherits only one gene, then he or she doesn't have sickle-cell anemia but is said to have the sickle-cell trait. People with this trait aren't especially ill, because the malformation of their red blood cells is not serious. They may, however, tend to get tired somewhat more easily than people without this trait.

So far, it doesn't look like sickle-cell genes have any more right to exist than do schizophrenia genes. At best, they can make people prone to tiredness. At worst, they can eventually kill the person who has them. But they do have a surprising benefit.

Sickle-cell anemia is particularly prevalent in parts of Africa where malaria is also present. (It spread to other places because of emigration and the mixing of populations.) Malaria, another terrible disease, is caused by a parasite rather than by one's genes. This parasite spends a part of its life cycle in a host's red blood cells. Fortunately for people with the sickle-cell trait, their mutated red blood cells are inhospitable to the malaria parasite, making them resistant to malaria. This makes the sickle-cell trait a lifesaver in those parts of the world where malaria is still a major public health issue.

There is no advantage at all to having two sickle-cell genes, but there is a benefit to having only one. Most of the people who inherit a sickle-cell gene inherit just one, rather than two. This mutation therefore helps more people than it hurts (at least where malaria is present), which is probably why Darwinian natural selection hasn't eliminated sickle-cell genes from the human genome. They're worth keeping around.

It's been proposed that schizophrenia's evolutionary origins and persistence across generations result from a similar mechanism. The full disease has no upside. However, a diluted dose of these genes can confer an advantage.

People with schizophrenia genes that haven't yet been fully activated by the environment (John Nash before his illness took hold), as well as people who have inherited a smaller number of these genes (speculatively, Ludwig van Beethoven), are not considered schizophrenic. The same is true of apparently healthy blood relatives of schizophrenics. Many such people are "schizotypes." They aren't mentally ill, though they may have watered-down versions of some of the unusual thought patterns that schizophrenics have, including—you guessed it—the tendency to think with remote associations.

Schizotypes may often seem unconventional, slightly odd, and sometimes even brilliant. They make connections that others don't, seeing the world in their own way.

You don't have to be a schizotype to be creative. However, schizotypes' flexible and creative thought processes give them an advantage. Cognitive neuroscientist Anna Abraham and colleagues posed a series of insight and analytic problems to schizotypes and non-schizotypes. The schizotypes solved more of the insight problems and just as many analytic problems compared to non-schizotypes. Schizotypes are more insightful. Other studies have shown that when schizotypes are given pairs of words and asked to judge the relatedness of the words in each pair, they rate the words to be more similar than non-schizotypes do. This demonstrates the kind of inclusive thinking that allows them to see obscure connections. It's also likely why healthy blood relatives of schizophrenics—those who have some schizophrenia genes—tend toward creative professions more often than people who aren't related to a schizophrenic.

Because it's written in your genes, schizotypy is a relatively stable personality trait. Though a schizotype may become a little more or less schizotypic over time as the environment activates or deactivates some of the relevant genes, these kinds of changes would usually be gradual. You probably can't think of any creative types who suddenly decided that being an accountant is their life's calling. And few accountants suddenly quit to become poets. These things can happen, but you don't see them every day.

GETTING IT ON AND GETTING ALONG

For this evolutionary model of schizophrenia to be a viable explanation of creativity, it's not enough that schizotypy confers a cognitive advantage. Humans are cognitively superior to cockroaches, but from the Darwinian standpoint, these pests are no less successful than we are. And even if a cognitive advantage does help people to live longer, that doesn't guarantee that their genes will be passed down. Staying alive isn't enough. Evolutionary success means reproducing more, not living longer. For schizophrenia genes to be passed down through history, carriers of these genes must, on average, have more offspring than do people who aren't carriers, or, in the modern era of birth control, they must at least have more sex.

But schizophrenics have fewer, not more, children than non-schizophrenics. This contradicts the idea that schizophrenia is passed down because it confers an advantage.

But what about schizotypes? Does their mildly unusual behavior inhibit their procreation? Apparently not. Schizotypes tend to have more sexual partners than non-schizotypes. This supports the idea that a small dose of schizophrenia genes makes one more likely to procreate.

However, schizotypy isn't a monolithic personality trait. It consists of several components, two of which are related to schizotypes' reproductive proclivities. The "unusual experiences" component contributes to one's creative output (for example, poems, paintings, songs, and so forth), which may serve as a kind of display to attract mates in the way that birdsongs do. (It's not lost on teenagers that one of the best ways to find romance is to join a rock band.) Another is "impulsive nonconformity," which describes people who don't conform to social norms or expectations and who are low in impulse control.

These characteristics can make schizotypes seem odd, sometimes to the point where they cause discomfort or discord. Nevertheless, many of the best managers and leaders in business, education, and other walks of life realize that creative ideas tend to come from unusual people and that if an organization wants to be innovative, then it must accept their quirkiness. However, the boundary between schizotypy and schizophrenia can sometimes be thin. It can be difficult for extremely impulsive nonconformists to work productively with others. As Ed Catmull, president of Walt Disney Animation Studios and Pixar Animation Studios, said, "There's very high tolerance for eccentricity; there are some people who are very much out there, very creative, to the point where some are strange." He values that creative eccentricity and is willing to tolerate a lot of the weirdness that often accompanies it. But movies are made by teams of people and not by a single person, so he has to draw a line. "There are a small number of people who are, I would say, socially dysfunctional, very creative," he said. "We get rid of them."

UNINHIBITED THOUGHT

Impulsive nonconformity doesn't cause creativity. Creativity and impulsiveness are both products of something else. This "X-factor" is a lack of inhibition.

The reduced inhibition at the heart of impulsive nonconformity is more than an inability to restrain one's desires. Inhibition, as a cognitive psychologist thinks of it, regulates emotion, thought, and attention. It's a basic property of the brain. Here's an example. Imagine that you are looking at a series of large numerals composed of small characters, such as the following:

Your task is to name each large numeral as quickly as you can while ignoring the small characters. In this case, you would say "seven" as quickly as you can. That was easy. Now say the next large numeral as quickly as you can:

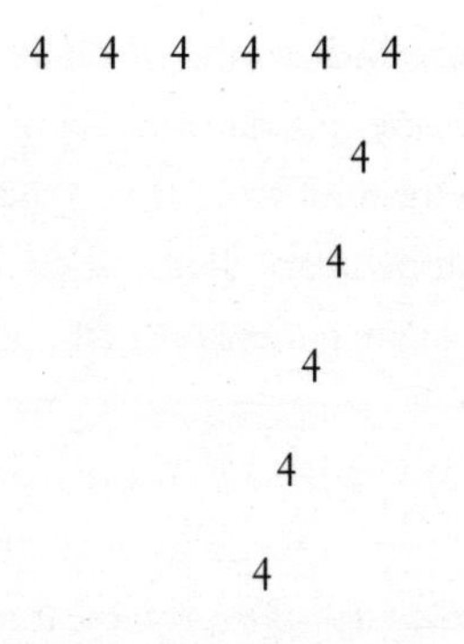

On average, you would say this "seven" a little bit slower than the last one, because the little 4s reflexively trigger a response ("four") different from the correct one ("seven"). These two tendencies compete with each other to be the actual response. To respond with "seven," you need a little extra time to inhibit your tendency to say "four." Next, try this one:

You'd think that it would be just as easy to say "four" to this 4 made out of #s as it was to say "seven" to the 7 made out of #s. But if you saw these numerals in this sequence, it wouldn't be. When you saw the previous 7 made out of little 4s, your brain had to inhibit the tendency to say "four." This inhibition isn't permanent, but in most people it doesn't dissipate immediately. So when you subsequently saw a big 4, the inhibition of the little 4s that came before slowed down your processing of the large 4 that followed.

More generally, when you purposely ignore something, even briefly, it's difficult to immediately shift mental gears and pay full attention to it, a phenomenon called "negative priming." This can sometimes be a minor inconvenience, but it occurs for a reason. When you ignore something, it's because you deemed it to be unimportant. By inhibiting something that you've already labeled as irrelevant, you don't have to waste time or energy reconsidering it. More generally, inhibition protects you from unimportant, distracting stimuli. Without it, your thoughts would wander all over the place, even retracing previous mental steps. Not coincidentally, schizotypes show little or no negative priming. Their reduced inhibition makes them susceptible to distracting, odd thoughts but also empowers their creative leaps.

Does all this mean that we are completely locked into whatever type of thinking our genes dictate? Is our DNA our destiny?

No. The person with the longest legs doesn't always win the race. The smartest kid in the class doesn't always get the best grades. Long legs and natural intelligence are certainly helpful, but these aren't the only reasons people win races or ace exams.

People differ in insightfulness, but these differences aren't set in stone. There is plenty of room to influence them. We've already shown how a positive mood can enhance insight by expanding the range of possibilities that you are willing to entertain. A recent study even suggests that alcohol can enhance insight by reducing inhibition and broadening attention, as long as you don't overdo it! So inhibition isn't completely fixed. You don't have to be a schizotype to broaden your mind. Nor do you have to intervene with a drink or a comedy video, though those interventions can have their own rewards.

In fact, keeping an eye on the clock can sometimes be enough.

YOUR FINEST HOUR

For most of us, levels of inhibition vary over the course of a day. Some people are "night owls" who are at the top of their game in the evening. Some are "early birds" or "larks" who peak in the morning. Others are in a slow transition from one type to the other. Many senior citizens tend to be up with the chickens; young adults are often creatures of the night.

People have their maximum inhibitory power at their peak time of day, whenever that is. This means that most young people are at their lowest ebb of self-control and focus in the morning; for most elderly people, depleted control and focus come in the evening. Things are actually somewhat more complicated than this because a person's accumulated sleep debt also has a major impact on inhibition. For example, teenagers are often more sleep deprived than adults, which saps teens' inhibition. However, all things being equal, you are at your inhibitory best during your peak time. This means that you are most analytical at your peak time and most insightful at your off-peak time. For creativity, your finest hour is literally the low point of your day.

SELF-CONTROL

The fact that the time of day influences the way you think shows that your cognitive style isn't completely fixed. This is of limited practical use because you can't control the clock. But you do have some control over how your attention is focused, as cognitive neuroscientist Ezra Wegbreit showed in Mark's laboratory.

Imagine that you are a participant in Wegbreit's study. It's a version of the flanker task, which we mentioned earlier. You see a row of three letters. Your job is to quickly press one of two buttons to report whether the middle one is an "S" or a "T." The first and third letters in each row can be ignored. However, the first and third letters can also be an "S" or a "T":

STS

or

TST

In these examples, the flanking letters incite the tendency to press the wrong button, causing an internal conflict that requires a bit of extra time to resolve before you can press the correct button.

But there's a strategy that can help you to minimize this conflict: You can try to ignore the flankers by narrowing your attention and focusing it on the middle letter. Some people are better at this than others, but just about everyone can sharpen his or her focus enough to reduce the interference. Wegbreit used this task to narrow people's attention for a while. At other times, he used another task to broaden it.

Going to the movies can be an epic experience. The screens in movie theaters are often so huge that if you sit close to the screen, you can't take in the whole picture at once without turning your head. But if you move a few rows back, with some mental adjustment, you are able to take in the big picture.

Wegbreit was able to reproduce this kind of voluntary broadening of attention in the laboratory. His participants were asked to identify the animals in pictures that were flashed on a computer screen for only forty milliseconds each. Because each picture was displayed for so little time, the participants didn't have enough time to scan around and note individual visual details such as ears and wings. They had to expand their attention to take in each picture in one all-encompassing glance. With a little effort, people can do this.

People can voluntarily broaden or narrow their attention. This isn't news. Wegbreit's innovation was to show that these adjustments can influence one's cognitive style.

After the participants' attention was either broadened or narrowed, they tackled a set of remote associates problems. Narrowing increased the number of analytic solutions; broadening increased insights.

So you do have some voluntary control over your mental spotlight, and you can use this to optimize your problem solving. But Wegbreit's study demonstrates another important point: This control is sluggish. Once your attention has been broadened or narrowed, it stays that way for a while. It doesn't immediately snap back to its previous focus. Your attention and cognitive style are flexible, but only within limits.

WEIGHTY INSIGHTS

We think that the insight-analytic dimension is something like a person's weight. Weight tends to gravitate toward a person's genetically influenced "set point." Partially inherited, your set point is also influenced by your present environment and past experience. Think of all the high-calorie food readily available within a few miles of your home. This kind of environment can shift your set point higher. Exercise, low stress, good nutrition, and ample sleep can lower it. We surmise that each person has a cognitive set point somewhere on the insight-analytic dimension. There are things that can push your set point in one direction or the other. However, genetics is still the gravitational force that will tend, without an intervention, to pull your cognitive style toward your original set point. The power of genetics was demonstrated one day in Mark's lab.

Over the years, we've tested many hundreds of people with remote associates problems. Of all the participants we've tested, one of the worst performers took part in an experiment in which participants were given thirty seconds to solve each of 144 problems. In this study, whenever a participant solved a problem, whether after five seconds or twenty, he proceeded to the next one. But when he didn't solve a problem, then the trial dragged on for the whole thirty seconds. With poor performance, the experiment could take a painfully long time.

One day, the experimenter sat in frustration as a participant failed to solve problem after problem. He seemed to be giving it his best effort, but unsolved problems were useless to us, so the experimenter considered stopping the session halfway through and cutting him loose.

Then, a funny thing happened. A particularly tough problem came up: "field"/"cry"/"ship." Only 18 percent of our participants were able to solve this problem within thirty seconds. The experimenter thought *He'll never get this one.* But very quickly the participant gave the correct solution: "battle." Surprised, the experimenter decided to press on, but the participant returned to his poor level of performance. In the end, he solved only 9 percent of the problems. The average for the other participants was more than 50 percent.

Two weeks later, the same experimenter was preparing another test session when a familiar-looking person walked in. "Have you done this experiment before?" asked the experimenter. He swore that he hadn't. The session started out like a repeat of the horrible performance of the other participant, and the experimenter now recognized where he'd seen the familiar face. "Are you *sure* you haven't done this before?" Again, the answer was "No, I haven't." Halfway through another agonizing session full of unsolved problems, the familiar-looking participant encountered the difficult "field"/"cry"/"ship" problem—and solved it. By this time, the experimenter was certain it was the same person whom he had tested a couple of weeks earlier. He suspected that the fellow was trying to get course credit for participating in the same experiment twice. But he didn't want to confront the student and he had some time to waste before the next participant arrived, so he decided to let the student complete the session. Once he left, the experimenter looked up the name of the prior participant, and was shocked to discover that it was a different person—with the same last name. The two participants were identical twins (who are, of course, genetically identical), and each performed remarkably similarly to the other. Both solved less than 10 percent of the problems, but both solved the "field"/"cry"/"ship" puzzle.

Experience and the environment are important, to a point.

12

CARROTS AND STICKS

Innovation has nothing to do with how many R&D dollars you have. . . . It's not about money. It's about the people you have, how you're led, and how much you get it.

—Steve Jobs, co-founder of Apple, Inc.,
quoted in *Fortune* magazine, November 9, 1998

Insightfulness may be rooted in genetics, but that doesn't mean that it's fixed. Whether you're an Analyst or an Insightful, your DNA doesn't totally lock you into only one way of thinking about the world. Psychological traits almost always involve a dance between one's genes and the environment, so your circumstances have a say. The stories of Judah Folkman, Hermann von Helmholtz, and Jerry Swartz show how your surroundings can enhance your creativity by improving your mood. Even annoying turns, such as Erik Verlinde's forcibly extended vacation, can prepare your mind for an aha moment by destabilizing unproductive thought patterns through fixation forgetting. As we'll see, your environment can tweak your cognitive style in other ways.

THE PARADOX OF MOTIVATION

Let's say you are a teacher trying to induce your students to write imaginative essays. Or perhaps you are a CEO intent on increasing your employees' creativity so that your company can be more innovative. If you ask an economist how to make these things happen, she will probably tell you to incentivize creativity either by rewarding it, punishing its absence, or both. A CEO could offer employees financial incentives, such as bonuses or stock options, for novel ideas, or terminate them for a lack of innovation. A teacher could give his students high grades for creative effort and low ones for a lack of it. After all, people will do more of almost anything if they are properly motivated. Won't they?

Yes and no. Yes, when offered incentives, people will certainly try harder to come up with ideas. No, because the result may be less creative than if the incentives had not been offered. Sam Glucksberg demonstrated this in a famous 1962 study.

Glucksberg asked participants to solve the classic Candle Problem. Each person was given a thumbtack box, about fifty thumbtacks, a matchbox, about thirty-five matches, and one wax candle. Their task was to mount the candle on the wall.

What won't work is softening the side of the candle with a lit match and sticking it directly to the wall. The candle is too heavy and falls off. And if you try to tack the candle directly to the wall, the wax crumbles.

There is, however, a way: Take the thumbtack box, empty it if there are any tacks in it, tack the box to the wall, and then mount the candle on top of it. To stabilize the candle, first light a match and heat the wax on the bottom of the candle to soften it so that it will stick to the box (see figure 12.1).

Glucksberg used two slightly different versions of this problem. One group of participants saw the thumbtacks presented inside the thumbtack box. For the other group, each box was empty, and the thumbtacks were loose rather than in a container.

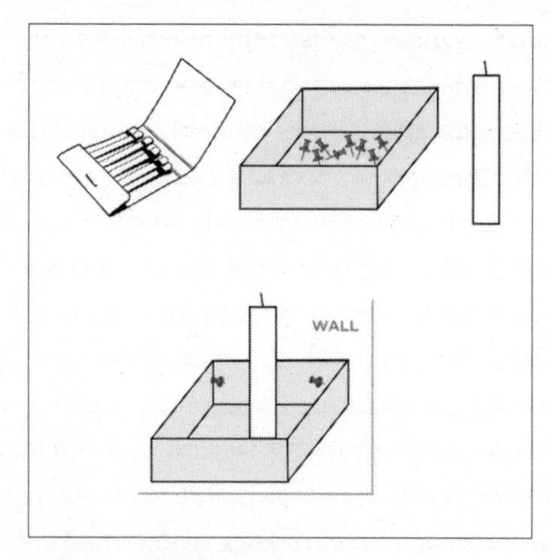

FIGURE 12.1: The Candle Problem.

When the tacks are presented in the box, this induces "functional fixedness"—it's hard to think of the box as anything other than a container. That's why, as we've seen, solving the problem usually requires an insight to change one's perspective about what the box is for. Few people solved this version of the problem, and those who did took a relatively long time. However, when the tack box is empty, its container function isn't prominent, and a change of perspective isn't necessary to think of it as a platform for the candle. People solved the empty-box version relatively quickly, easily, and analytically.

So far, this is a typical insight experiment.

The novel part of this study was the financial incentives. Half the participants were just asked to solve the problem. The others were told, "Depending on how quickly you solve the problem, you can win five dollars or twenty dollars." That's 1962 dollars—real money for these college-student participants.

In the analytic, empty-box version, offers of prize money induced more people to solve the problem. But in the insight, container-box version, incentives actually reduced the number of people who were able to solve the problem; money prizes also increased the amount of time that the few successful solvers needed.

Paradoxically, stronger motivation decreases insightfulness. To find out why, grab a bag of peanuts and head to a park.

ZOOMING IN AND LOCKING ON

When you see a squirrel, offer it a peanut. At first, the squirrel will start coming toward you to take the treat. But as it gets closer, it will become afraid and, as its fear starts to get the upper hand, it will slow down and eventually stop. It may back up a little, then come closer again, back up, and so forth.

Scientists have discovered two opposing motivational systems in the brain. The "approach system" draws you toward a goal, such as an appealing dessert you want to eat or an annoying person you want to confront. The "avoidance system" pushes you away from things like a threatening weapon or rotting food. In the squirrel's case, the wrestling match between the approach and avoidance systems may go on for some time before one gets the upper hand: either the squirrel inches up to you and takes the peanut or it gives up and looks elsewhere for a less-threatening meal (which you shouldn't take personally).

Even though these two systems drive your behavior in different directions, they have the same effect on your attention: they lock your mental spotlight on the object that you crave, fear, or loathe, and then they narrow the beam. The squirrel sometimes creeps forward, sometimes backward. But all the time it's focused like a laser on you and the peanut.

Cognitive neuroscientist Eddie Harmon-Jones and colleagues demonstrated this in the lab by showing a series of pictures to their participants. Some pictures—delicious desserts—engaged the approach-motivation system. Disgusting pictures, such as rotting food, activated the avoidance system. Other pictures showed motivationally neutral objects, such as a table, that didn't involve approach or avoidance.

After each picture, each person's attention was measured to see if it was broadened to view the "forest" or narrowed to see the "trees." This was done by showing them a large letter composed of small letters. For example:

M
M
M
M
M M M M

The participants' task was to quickly press a button to show that a particular target letter was present either as the big letter or one of the small ones. By measuring how quickly participants responded to target letters when the letters were large or small, the researchers could assess whether their attention was deployed broadly or narrowly. Both desserts and disgusting pictures narrowed people's focus (relative to neutral pictures).

These findings suggest that both enticements and threats should degrade creative insight because they constrict attention. But other research reveals a subtler story.

THE APPROACHING PARADOX

You are a mouse navigating your way through a maze. For a moment, you look upward and see a hungry owl flying overhead and gazing down at you. You redouble your efforts to find the exit that leads to a safe haven. After scurrying around, with an occasional glance upward, you finally discover the exit and step through it into a protected compartment.

How do you feel?

Social psychologists Jens Förster and Ronald Friedman used this scenario to induce avoidance motivation in people. Participants read the scenario and then traced out an exit route on a paper-and-pencil maze. In this case, solving the maze narrowed their attention and impaired their insightfulness.

Now consider an alternate, less scary scenario, sans predators. You are a hungry mouse searching for the maze's exit because you know that it leads to a compartment with a piece of cheese. You find the exit and consume your prize.

Now how do you feel?

You might expect that solving the cheese maze would constrict attention in the same way as viewing pictures of desserts. However, the researchers found that the cheese maze broadened attention and enhanced insightfulness.

This wasn't a fluke. It seems that approach motivation has paradoxical effects, sometimes broadening the mind, as in the case of the cheese, and sometimes narrowing it, for example, when prize money is offered. There is, however, an overarching principle that resolves this paradox and explains how motivation influences your creativity.

TYPES OF MOTIVATION

The key is that there are different kinds of motivation. One is about specific objects of personal significance: things such as cheese and owls draw you in or repel you, grabbing and narrowing your attention in the process.

Another type of motivation is about you rather than about the objects around you. It comes in two flavors: prevention and promotion. As social psychologist E. Tory Higgins explains, these describe how you relate to the world.

When you are oriented toward prevention, your central concerns are safety, responsibility, and the avoidance of pain or loss. ("I don't want him to fire me." "I'm playing it safe.") A prevention mind-set banishes long-shot ideas and remote associations because they are risky. It narrows and focuses your mind on the most obvious features of your situation and recruits analytic thought to methodically evaluate what seems to be most important. It sticks with the tried and true. That's why your attention is still constricted even after you've successfully evaded the owl—fleeing it has left you in a prevention mode.

In contrast, a promotion orientation permits you to take risks because it starts from the presumption of safety. Since you believe yourself to be safe, you feel free to look at the world in terms of your hopes and possibilities for advancement. Promotion orients you to embrace creative, untested means of achieving your larger ambitions. ("I want to be something!") And when you are oriented toward achievement, advancement, and progress, you experience a sense of adventure empowered by a feeling of safety. *Dream big. Everything is possible.* This broadens your mind to the universe of opportunities. It enhances and endorses creative leaps that can help you to achieve your vision.

This resolves the paradox. Approach motivation constricts the mind and stimulates analytic thought when it's about a concrete object that you want. However, it broadens the mind and energizes creative insight when your motivation is a general drive for self-enhancement.

PRIZES AND PINK SLIPS

The relationship between motivation and attention has unexpected implications that contradict some aspects of traditional economics. Consider the XPRIZE Foundation. Its goal is to stimulate innovation by offering $10 million awards to the winners of its competitions. According to its website, "the XPRIZE Foundation is an educational nonprofit organization whose mission is to create radical breakthroughs for the benefit of humanity thereby inspiring the formation of new industries, jobs and the revitalization of markets that are currently stuck. Today, it is widely recognized as the leader in fostering innovation through competition." One of its awards, intended to advance commercial spaceflight, went to a team led by aerospace designer Burt Rutan and Microsoft co-founder Paul Allen for designing, building, and launching (twice within a two-week period) a "spaceplane" that carried three people one hundred kilometers above the earth's surface. Another competition will reward the developer of a new technology for quickly and economically sequencing DNA.

Clearly, big prizes can have benefits. If a person already has a breakthrough idea before the announcement of a contest, then the allure of the prize may energize him to mobilize a team to work out the details and implement the idea. Contests can also help to publicize important technological and scientific goals and recruit a large number of people to work on a problem. All things being equal, there is a greater chance of solving a problem if there are more minds grappling with it.

But all things are rarely equal. We've seen how prizes have the power to constrict creative thought and inhibit innovation. The unanswered question is whether these thought-constricting effects are long lasting. Even if a team works on a project for months or years, the allure of the prize may not recede from awareness enough to loosen its grip on attention to allow flashes of creative insight. The award would increase the odds of a creative breakthrough only if its stifling of creativity were offset by a large enough increase in the number of people working on the problem.

Another example of extreme rewards and punishments in the business world is Jack Welch's strategy when he was CEO of General Electric (1981–2001). Welch lavished stock options and bonuses on the top 20 percent of his managers. He also fired the bottom 10 percent of his managers annually. This policy may have enhanced managers' analytic thought and general work output, but it's unlikely to have contributed to the kind of consistent innovation that a technology company needs for long-term viability in the high-tech arena. Unsurprisingly, during Welch's tenure as CEO, 40 percent of General Electric's earnings came from GE Capital, the financial services division of the company, rather than from the more-famous technology side of its business.

READY OR NOT

Another common feature of pressure-cooker companies, schools, and contests is the deadline. When not carried to extremes, deadlines can be useful for motivating people and focusing them on the task at hand. But what happens when the desired outcome is a new idea or a solution to a problem? If you're in an analytic mode of thought and you run out of time, then you can always submit your work in progress: the idea that you were consciously considering when the clock ran out. This idea may or may not be good enough. Insightful thought, on the other hand, is mostly unconscious and offers up a solution only when complete. If you're in an insight mind-set, you won't have anything to offer until you've had your aha moment. That's why, as we've seen, when confronted with a deadline an Insightful tends to make "errors of omission" by drawing a blank, while an Analyst's errors are more likely to be incorrect solutions ("errors of commission").

Deadlines can have an additional negative effect on creative work. The threat of punishment or loss associated with missing a deadline can cause anxiety and avoidance motivation, which push you into an analytic mind-set. Of course, moderate anxiety can motivate you to produce something by the deadline. But if your goal is a creative product, then the threat of being late will make it that much harder for you to innovate.

Consider the process by which U.S. scientists typically obtain grant funding to pursue their research. Scientists prepare detailed research proposals for submission to funding agencies such as the National Institutes of Health (NIH) and the National Science Foundation (NSF). The NIH has three grant submission deadlines per year; the NSF has two. If a scientist misses one of these deadlines, she will then have to wait for several months to submit the application. If the result of this delay is an interruption in her laboratory's funding, then the consequences can be devastating. Lab personnel may have to be terminated. The lab may even have to shut down. In many cases, part or all of the lead researcher's salary is supplied by such grants, so even a brief blip in the flow of funding can result in a substantial salary reduction or even the loss of a job. This creates enormous pressure to submit a regular stream of grant proposals and to submit them promptly for every deadline.

In the United States, there is a great deal of concern—fueled by recent evidence of decreasing creativity test scores—over what some believe to be dwindling innovation. Of course, everyone wants government to be efficient. The system of infrequent grant proposal deadlines helps government agencies process proposals effectively. Unfortunately, it likely also hinders innovation. The obvious fix would be to eliminate the deadlines, or at least make them much more frequent. Researchers would be more likely to relax and wait for creative ideas to flow if they knew that their proposals would be evaluated soon after submission, no matter when they were submitted.

Money can't buy love, but it can certainly attract suitors. Similarly, money can lure people to work on particular problems, but it can't make them have breakthrough ideas. Prospects of financial gains and losses don't have the same power to enhance or suppress creative thought that they have over noncreative behaviors. There is, however, another surprising lesson that will dumbfound many economists. Since creativity flows from a promotion orientation, not prevention, creative insight will be enhanced *after* a prize is won, not in anticipation of one. So make your way through the maze and eat your cheese in safety. Then your mind will expand.

13

FAR, DIFFERENT, UNREAL, CREATIVE

Toto, I've a feeling we're not in Kansas anymore.

—Judy Garland as Dorothy Gale in *The Wizard of Oz*

What's it like to be a punk? Assuming that you weren't ever one, take a minute to consider this question. It will change you, at least for a while.

We've seen how moods and motivations can shift your cognitive style to make you more insightful or analytical. However, these kinds of "hot" emotional processes aren't the only influences on creativity. "Cold" cognitive processes can also have a potent impact. Thinking, as well as feeling, can energize the mental machinery that produces creative insights.

SYMBOLS

Jens Förster, Ronald Friedman, and collaborators theorized that high creativity is associated in people's minds with being different from the norm. Any symbol of deviance (in the statistical sense) should therefore prime thought patterns related to creativity. To test this idea, they asked participants to spend a few minutes writing about what they thought it would be like to be a punk (which generally doesn't have a negative connotation in Germany, where this study was done); they asked others to write about what it would be like to be an engineer, which in Germany is thought to exemplify conformity. Their participants rated punks and engineers as equally creative, interesting, and likable. Nevertheless, those asked to think about punks solved more insight problems than those asked to think about engineers. Participants asked to think about engineers solved more analytical problems. The researchers' conclusion: Thinking about unusual people primes creativity; thinking about conformism brings out one's analytical side.

In another appeal to people's conception of deviance, the researchers brought other participants to a laboratory to work on a set of insight and analytic problems. A poster of an abstract picture was displayed in the laboratory as a decoration. It was clearly visible, but the experimenter didn't call participants' attention to it. Each person saw one of two versions of the poster (see figure 13.1).

The participants who saw the deviance poster scored better than

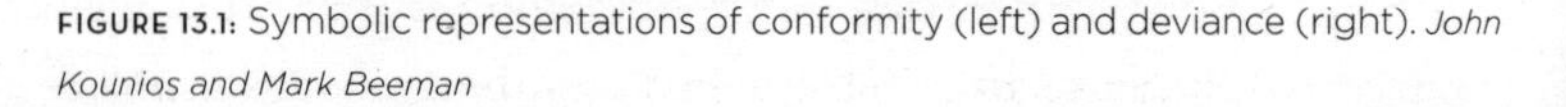

FIGURE 13.1: Symbolic representations of conformity (left) and deviance (right). *John Kounios and Mark Beeman*

those who saw the conformity poster on a test of creativity requiring them to think of new uses for common objects. This subtle manipulation of the environment was enough to influence their creative behavior.

If symbols of deviance can prime insightful thought, it follows that symbols of creativity would have the same effect. In one example, participants arrived at the lab of social psychologist Michael Slepian to work on insight and analytic problems. Before beginning, the experimenter commented that the room was a little dark, at which point he turned on a light. For some participants, the light was fluorescent; for others, it was a bare incandescent bulb. The participants who saw the incandescent bulb switched on solved more insight problems but didn't solve more analytic problems; however, the fluorescent light failed to boost performance for either type of problem. Only the classic symbol of creativity spurred insight.

It's reassuring that scientific studies have verified the idea that such symbols have the power to inspire creative insight. However, this isn't particularly surprising. People have found creative inspiration in them for millennia. Other research has gone further to demonstrate more unexpected ways in which one's circumstances can prime insightfulness.

LOVE AND THE TIME TRAVELER

Suppose you won a trip to Tokyo but were told that you had to embark on your journey soon—tomorrow! After you got over your initial surprise, what would you think about? *Is my passport still valid? Yes, I believe so. What clothes will I need? I'm going to have to do laundry tonight. Should I take a taxi to the airport, or can I get someone to drive me? What time does my flight leave? I'm going to have to reschedule that appointment with my dentist.* And so forth.

Now, instead, let's suppose that your trip to Tokyo would occur one year from tomorrow. What would you think about? *It will be exciting to visit another country. What can I do during this trip? How should I prepare? I'm going to research places to visit that will interest me. I will learn a lot. I'll read up on Japanese culture and history. I hope I meet some fascinating people. I should try to travel a little more often. Multiculturalism is enriching.*

If you imagine taking a trip tomorrow, you'll tend to think about specific details of the trip. But if you imagine a trip in the more distant future, you will expand your scope to think in more abstract terms about its meaning and purpose. These two types of "mental time traveling," also known as "prospection," have contrasting effects on your cognitive style. Near-future mental time traveling gives a temporary boost to analytic problem-solving ability. Distant-future time traveling enhances insight.

It seems natural to assume that creative leaders think about long-term consequences because they are creative and that their less-creative counterparts tend to think about short-term consequences because they are less creative. To an extent, there is probably something to this idea. But causality apparently flows in the other direction as well: Considering long-term consequences enhances creativity.

Förster and his colleagues went on to extend their findings in a surprising way. They argued that romantic love transforms one into a mental time traveler. The thoughts of a person in love often float to the distant future as he or she imagines a lifetime with the beloved. But when a person lusts after another without romantic feelings, then he or she becomes a mere operator, focusing only on the mechanics of the real or imagined short-term possibilities. The researchers primed some participants by asking them to think about someone for whom they have felt romantic love and primed others by asking them to think about someone to whom they were sexually attracted without romantic feelings. The romantics performed better on a set of classic insight problems but not on analytic problems; the lusty participants did better on the analytic problems. A follow-up study primed participants by flashing love- or lust-related words on a computer monitor so quickly that the participants couldn't consciously identify the words. Again, they found that love priming enhances creative thought, while lust priming triggers analysis. Participants' attentional focus was also measured: As predicted, surreptitious love priming broadened attention; lust priming narrowed it.

The researchers proposed that love and lust influence insight and analysis by eliciting thoughts about the distant or near future. But we think that additional factors are at play. People may be happier when in love than in lust. The elation of love may enhance creativity in the same basic way that watching comedy videos did in our research, by putting one in a positive mood. Moreover, when in love, a person tends to take the perspective of his or her beloved. He or she thinks about the beloved's likes, dislikes, hopes, and dreams. Love therefore demands the cognitive flexibility to change perspective. Lust, however, is not accommodating. It is inherently inflexible and self-oriented.

There are other forms of mental travel. Besides traveling to worlds that don't yet exist, one can also travel to worlds that will never exist.

WHAT IF . . . ?

How would your life be different right now if your parents had won $100 million in a lottery when you were a child? What if the Soviet Union had never dissolved? What would the world be like if all disease were eliminated?

These are examples of "counterfactual" thinking—that is, thinking about scenarios that contradict reality. "Additive" counterfactuals add something to a situation to create a new scenario: "If I had an umbrella, then I wouldn't have gotten soaked." Subtractive counterfactuals take something away: "If it hadn't rained today, then I wouldn't have gotten soaked." Social psychologist Adam Galinsky and his collaborators primed people by asking them to think about such alternate worlds. Curiously, they found that imagining a new world by adding something enhances creative thought; taking something away improves analytic problem solving.

We suspect that these two types of counterfactuals have different effects on cognitive style because they have contrasting influences on attention. To entertain an additive counterfactual, you have to expand your awareness beyond your current situation to include extra things in the mix. It's the here and now *plus* something. This should broaden conceptual attention and enhance insight. But when you consider a subtractive counterfactual, you have to think about the here and now *minus* something. This isn't as simple as pushing the banished thing out of your mind. Suppose someone tells you, "Don't think about a pink elephant." Can you do it? Not so easy. The beast can't just be erased. Instead, you have to narrow your conceptual attention and focus on something else to exclude the pink elephant from consciousness. This constriction of awareness may be why subtractive counterfactual thinking enhances analysis.

If merely thinking about alternate realities can change how you think, it should come as no surprise that actually living an alternate reality can have the same effect.

CREATIVITY FROM AFAR

Many great artists (for example, Paul Gauguin and Pablo Picasso), musicians (George Frideric Handel and Igor Stravinsky), and writers (Ernest Hemingway and Vladimir Nabokov) did much of their best work either while living in foreign countries or right after. Creativity seems to be high in first- and second-generation immigrants (Nikola Tesla and Thomas Edison) as well. Adam Galinsky and his collaborators investigated this apparent link between multiculturalism and creativity by asking their participants to report on how much time they spent abroad and by asking them to take tests of creativity, including several types of insight problems. There was no relationship between these measures of creativity and the amount of time they had spent traveling abroad. But there was a positive association between insight and the amount of time spent *living* abroad, the critical distinction being whether a person had to adapt to another culture rather than just skim along its surface as a tourist. Furthermore, for those who had lived abroad, thinking about past experiences adapting to another culture provided an additional creativity boost.

Like love, adapting to another culture demands the cognitive flexibility to consider situations from another perspective. It involves learning to think about alternate, even opposite, interpretations of events and circumstances. (If you haven't ever lived abroad and don't believe this, just watch a foreign news channel or read some foreign newspapers to prove this to yourself.) It also requires broadened attention to keep your culture's viewpoints and those of the new country both in mind but separate so you don't confuse them.

The implications are straightforward. The multicultural effect explains why immigrants are such a rich source of creative talent and why many people have found the time-honored practice of studying abroad so beneficial.

We've seen that thinking about the distant future, faraway places, different cultures, and imaginary worlds can prime you for insightfulness. This is not a laundry list of unrelated techniques for enhancing creativity. These factors are all manifestations of a single psychological principle.

DISTANT THOUGHTS

Objects of thought are removed from you in time, space, social standing, and "counterfactuality" or "realness." Thinking about things that are psychologically close to you—places that are nearby, people who are like you, events that will happen soon, your immediate reality—narrows your attention, prompting you to focus on salient details and biasing you toward analytic thought. Thinking about places that are far away, people who are unlike you, things that will happen in the distant future, and scenarios that are different from your current reality all broaden attention and benefit creative insight.

This principle is actually quite profound. Hundreds of years ago, most people lived in small villages and rarely traveled far from home. They couldn't read, so they had little awareness of the larger world beyond their community. Now, travel is common, and advancing technology is fueling rapid globalization. Millions use Skype, Google Hangouts, and other Internet services to hold live teleconferences with people all over the world. People watch international television programs and videos on cable, satellite TV, and the Internet. When a consumer in the United States picks up the telephone to call customer service, she may well speak to someone in a call center in India. Physical and financial impediments to long-distance human interaction are thus dissolving. Globalization, it seems, will have an impact on humanity that is deeper than economic or cultural change. It may well be subtly shifting our cognitive style toward insightfulness, potentially slowing what seems to be an overall (U.S.) decline in creativity.

An example hints at the generality and impact of this principle. Social psychologist Lile Jia and his collaborators asked participants to solve a set of classic insight problems. One group of participants was told this test was developed at a research institute two miles away from Indiana University, where the study was done. Another group was told the institute that devised the test was located in California. The problems that the two groups saw were otherwise identical. Nevertheless, the participants who thought that the test was prepared in California solved more insight problems. Psychological distance, even when imaginary and incidental to a situation, enhances insightfulness.

The theory of psychological distance explains much, but some details remain to be worked out. It's not clear how it can explain the opposing effects of additive and subtractive counterfactual thought, both of which seem to involve psychological distance. Moreover, to our knowledge, it's not clear why psychological distance should prime creative thought only when measured with respect to oneself. (Why shouldn't the thought that the sun is far from the moon prime you to be insightful just as effectively as the thought that the sun is far from you?) These issues will likely be worked out. For now, though, the important point is that the effects of psychological distance on creative insight are real.

THE BIG PICTURE

When you're standing too close to a painting, you don't have the proper perspective for understanding it. Step back a few paces and you'll gain a new vantage point that allows you to see all of its parts and how they fit together to form a meaningful whole. This is the "thirty-thousand-foot view" or "Archimedean point" for taking in the whole situation. To achieve this, you must broaden your attention to see—and your mind to conceive of—all the interconnected pieces, even the seemingly trivial, nonobvious ones. Only a mind broad enough to take in all the parts can reconfigure them to make a new whole. Psychological distance primes this mental broadening.

Insightfulness often involves ramping up thought from the particular to the general, thereby raising the stakes to produce ideas whose usefulness and appeal go beyond the specifics of the original circumstances. Archimedes's specific task was to figure out whether the king's crown was made of pure gold. Looking at the big picture enabled him to solve this specific problem by realizing a general method for determining the volume of any object. Wag Dodge's goal was to escape from the wildfire. The narrow, literal sense of "escaping from the fire" was to run away from it. Broad thinking empowered Dodge to realize that escape doesn't just mean running away. It includes any method for creating physical space between the fire and him, including the possibility of making the fire run away from him. Isaac Newton observed a specific event—an apple falling from a tree. His expansive mind understood that this event represented the same universal force that shapes the motions of the planets. The theft of Erik Verlinde's valuables shook him out of his usual thought patterns so that he could leap beyond Newton's and Einstein's theories to reinterpret gravity as something yet more general and abstract: a by-product of the universe's tendency toward disorder.

So, to empower creative thought, you should imagine unusual people, alternate realities, and the distant future. Think about a loved one. Take in the view—and broaden it. Visit or envision faraway places and cultures. Ponder "what spring is like on Jupiter and Mars."

THE STATE

The empires of the future are the empires of the mind.

—Winston Churchill, speech at Harvard University, 1943

A very experienced Zen meditator once contacted John to ask to have his EEG recorded and compared with the EEGs of average nonmeditators. The idea was to demonstrate to his students the effects of long-term meditation. Curious to see if there was a difference, John agreed. After his EEG was recorded—and yes, his brain activity was unusual—the meditator asked about the lab's research. John told him about our work on the neural basis of insight, and he was intrigued. There was some time left in the session, so John asked him if he wanted to do a series of remote associates problems. He enthusiastically agreed.

John was sure that the meditator would blow these problems away. After all, he had a superbly honed mind and a doctorate to boot. However, after he attempted about a dozen problems, it looked like he was going to tank. He timed out on puzzle after puzzle without solving a single one. After watching a couple of dozen painful failures, John was about to stop the session to spare the Zen meditator further embarrassment. But then he got one right. And then another and another. From that point on, he got nearly all of them right.

This was unprecedented. Over the years, we have tested hundreds of participants and found no evidence that practice during a single session improves performance on remote associates problems. Nevertheless, the Zen meditator showed an amazing—and very sudden—improvement in his performance.

This was an informal observation rather than a formal experiment, so we can't offer a definitive explanation for the meditator's turnaround. We do, however, have an educated hypothesis. Years of intensive meditation may have granted our Zen subject with hyper-refined attention. His mind was so still, focused, and concentrated that associations didn't spontaneously pop into his awareness. However, his years of mental training also afforded him a great deal of self-control. Failing to solve these problems within his regular state of mind, he had the ability to try out different ways of allocating his attention until he hit on the broad focus that optimized his performance. Once he found the right spread of attention, his superior self-control allowed him to maintain it consistently, resulting in outstanding performance.

This wasn't our first brush with the idea that there may be practical strategies for enhancing people's insightfulness. In fact, whenever either of us meets someone new and the conversation turns to what we do for a living, we are usually asked for a bit more detail about our research. When we mention that we study how the brain has aha moments, the floodgates usually open wide. Our new acquaintances tell us about their own significant aha moments and ask us to explain exactly what happened in their brains when they had them. Soon enough, we are inevitably asked, "How can I have more?"

Everyone wants to be more insightful. But be forewarned: The Insightful faces obstacles.

Be careful what you ask for.

Jealousy, suspicion, and anxiety are common reactions to new ideas. People sometimes resist creative proposals simply because of a kind of prejudice toward their authors. ("Who does she think she is?") Highly creative people can be difficult to comprehend because they often leap over many steps of reasoning. ("His thinking is all over the place.") They can be seen as disruptive, even subversive, because they question established assumptions and procedures and seem distressingly ready to overturn the status quo. ("What's her *real* agenda?") Many highly creative people are schizotypes who are easy to view as odd and unpredictable. ("What's he been sniffing?") Moreover, many people don't want their comfortable worlds upended, even when there is a possibility of betterment. ("What we have is good enough. Doesn't she realize how risky that would be?") These kinds of reactions often incite people to block Insightfuls from taking leadership roles or prevent their suggestions from receiving a fair hearing.

In his autobiography, Benjamin Franklin discussed the many civic organizations with which he was involved, many of which he founded, such as the first fire department in Pennsylvania, the first public library in America, and the school that later became the University of Pennsylvania. Nowadays, it's hard to believe that many of the ideas Franklin proposed were received with skepticism and even outright resistance. He eventually realized that people are often jealous or suspicious of those who make creative proposals. His solution was to attribute these proposals to other people. When he did this, his colleagues were usually more receptive. Consequently, he didn't always receive his due credit, but he was willing to make that sacrifice to smooth the path of progress. (Franklin had already become so famous and wealthy for his many important ideas that this was no great loss for him. And he's made up for some of this lost credit. As a real estate broker once said, "People will accept your ideas much more readily if you tell them Benjamin Franklin said it first.")

But the risks that come with creativity can be dwarfed by its benefits. So let's proceed to consider three potential strategies for harnessing the eureka factor.

TRAINING THE MIND

The first approach involves mental training. There is no shortage of programs, books, and techniques to choose from. Some even claim to be "based on scientific principles." Unfortunately, the fact that there is little or no objective scientific evidence for the effectiveness of most of these regimens hasn't stopped business gurus and personal growth "experts" from hawking their products. This makes it hard to distinguish what works from what just seems good.

There have been only a handful of studies whose specific aim was to train people to solve problems insightfully. Their general goal tends to be to teach people to scrutinize the preconceptions that are the basis for their mental boxes. The idea is that by making yourself aware of your assumptions, you can systematically examine each one to ascertain whether it is correct or necessary. This will allow you to dismantle an incorrect perspective about a problem and construct a new understanding that will point to a solution.

You might wonder why you should consider this approach. Mental boxes are usually unconscious for a good reason. Constantly uncovering and evaluating your preconceptions would be exhausting, even for a professional philosopher. These preconceptions are the circuit breakers that protect your brain from overload. Doesn't constantly thinking about the box negate the purpose of the box?

Yes, but not always. It does make sense to scrutinize your assumptions when you're stuck on a problem. What are the individual pieces of information in the problem? How do these pieces fit together? How do they relate to other situations? What unnecessary assumptions are you making? Once you answer these questions you can deliberately put aside your understanding of the problem and construct a new perspective.

The underlying method is to compensate for the absence of insight by using analysis as a substitute. To accomplish this, a trainer presents a number of insight problems and explains how these problems typically lead to an impasse by coaxing a person to focus on misleading features that send him or her down the garden path. Awareness of these mental traps helps a person to avoid them and search for alternate interpretations. Consider an example.

Many years ago, people complained about the slowness of the elevators in a new high-rise building, so the building's owners brought in engineers to solve the problem. All of the engineers said the same thing: It would cost a great deal of money and wouldn't result in a noticeable increase in speed.

It seemed that this was a hopeless engineering problem. But another consultant offered a different kind of solution: Line the elevator cars and doors with mirrors. When the owners did this, people were so fascinated looking at themselves that they didn't notice the passage of time. End of complaints.

We don't know whether this consultant's reinterpretation of the engineering problem as a psychological problem was the product of an aha moment or the result of methodical evaluation of the problem's parts. But this is just the sort of insight problem that one can learn to solve analytically. You can do this by attending to all the parts of the problem rather than just the parts that seem at first to be most important. For example, people hear "The slowness of the elevators causes complaints" and think of this as "*The slowness of the elevators* causes complaints." But if you broaden your attention and note that *the slowness of the elevators* is just one piece of a problem that has two major pieces, then you will realize that the other piece might be the key. This empowers you to focus your attention on the other part so that the problem now becomes "The slowness of the elevators *causes complaints*." This simple shift of emphasis transforms an engineering problem into an easily solvable psychological one.

Now consider a difficult insight problem that most people aren't able to solve within a few minutes. You are given the following items: two steel rings, a long candle, a match, and a two-inch steel cube. Your job is to use these items to fasten together the steel rings in any way you like. Take a few minutes to try to solve this problem before you continue reading. But don't bother with the idea of fastening the rings together with melted candlewax. The rings are heavy, and the wax won't hold them together.

Functional fixedness is the impediment to realizing the solution: In this case, it's difficult to overcome the tendency to think of creating light as the sole function of the candle. Cognitive psychologist Tony McCaffrey recently developed a technique to help people avoid this roadblock. First, list and describe all the objects and their parts in a way that ignores their typical functions. You wouldn't include "candle" in the list, because "candle" implies the function of lighting. Instead, you would include "a long wax cylinder with a string embedded in its core" in the list. This neutral description enables you to think of the wick as something that can be used to bind the rings together. Then, the solution becomes obvious: You need to extract the wick from the candle. The only tool available to help accomplish that is the cube. You use the edge of the cube to scrape all the wax off of the wick so that you can use the wick as a string to tie the rings together. This neutral-description technique enabled participants to solve more functional fixedness problems than other people who were just told to solve the problems without being given any special instructions.

Clearly it's possible to improve a person's ability to solve some types of insight problems by training him to evaluate all the parts of a problem and scrutinize their assumptions. But these techniques strengthen analysis, not insight. When you use this approach, there is no aha moment. But in a pinch, it's often a reasonable substitute.

This same principle is at work when people learn to solve some types of insight problems analytically through sheer practice. In one of our recent studies, participants spent three sessions on different days working on a long series of remote associates problems. We found that they gradually made increasing use of analysis and decreasing use of insight over the course of the sessions. The experience of solving many highly similar puzzles evidently allowed them to develop effective analytic strategies. One example of such a strategy for solving remote associates problems is to think of several associations for the first word of the triplet and then systematically try out each of these associations on the second word and then the third word of the problem until the solution is discovered. Our participants were able to discover such strategies because all of these problems had the same basic structure. That's why a single step-by-step method could work for every problem. But experience with any single type of puzzle is unlikely to be of much help in solving other kinds of problems.

The key point: The right kind of training and experience may help you learn how to solve specific kinds of problems, but it won't help you solve other types of problems, and it won't teach you how to have insights.

Real-life problems aren't circumscribed. For an Insightful, there is no boundary between the problem and the rest of the world or the rest of one's experience. Young Judah Folkman passed his chemistry test by borrowing equipment from a physics lab. Isaac Newton conceived of gravity when he saw an apple fall from a tree.

Even puzzles aren't always clearly defined. The Nine-Dot Problem, which we described in Chapter 3, tasks one to draw four or fewer lines to connect nine dots arranged in a square without lifting the pencil from the paper. This literally requires outside-the-box thinking because, to accomplish this, one must extend the lines beyond the boundaries of the square. The difficulty lies in the fact that when looking at a square formed by three rows of three dots, people see the square as the problem—the whole problem. The solution therefore requires one to use something that doesn't seem to be a part of the problem. Insightfuls are good at this because they are able to draw ideas from anywhere and from anything. They aren't limited by written or unwritten rules. They are the street fighters of the mind.

Training has another limitation. As we've seen, insights are often spontaneous. You may not realize that you are stuck on a problem—or that you even have a problem—when a stray word or glance triggers an aha moment. Judah Folkman was thinking about blood substitutes, not cancer, when he suddenly realized that tumor growth depends on a supply of blood. Richard James wasn't thinking about designing a toy when the sight of a bouncing spring sparked his idea for the Slinky. Jerry Swartz was playing with his children when he suddenly conceived of a mechanism for a handheld laser bar code scanner. In contrast, the skill conferred by "insight training" is not spontaneous—you won't scrutinize your assumptions about a problem unless you realize that you have a problem. And if you try to scrutinize all of your assumptions all of the time, you will quickly exhaust yourself.

Moreover, some problems simply aren't amenable to analytic thought. When you don't have a hammer to pound in a nail, a rock can usually get the job done. But with a Phillips-head screw, nothing will substitute for a Phillips-head screwdriver. Sometimes, insight is the specific tool that you need. Unfortunately, it can't be pulled out of your mental toolbox at will.

PRODDING THE BRAIN

The second approach to enhancing creative insight is completely different. It involves directly stimulating areas of the brain involved in insightful thought. No, this isn't science fiction.

New techniques for intervening in the brain's functioning have recently entered researchers' armamentarium and are even being explored as therapeutic tools. One of these techniques, "transcranial direct current stimulation" (tDCS), has been used to try to facilitate insight. In tDCS, two electrodes are placed on the surface of a person's head, and a very mild electric current is passed from one to the other. (This is different from electroconvulsive "shock" therapy, which uses a strong current to forcibly reset brain activity.) Brain tissue near the "anode" electrode is excited; activity in tissue near the "cathode" electrode is suppressed.

Cognitive neuroscientists Richard Chi and Allan Snyder recently attempted to spur insight by using tDCS to stimulate participants' right hemispheres and inhibit their left hemispheres. Stimulation was applied while they tackled insight problems. (For comparison, other participants received "sham" treatment with the same equipment but no electrical current.) Stimulating the right hemisphere (while dampening the left hemisphere) substantially improved participants' problem solving compared with sham stimulation. Remarkably, it boosted performance on the classic Nine-Dot Problem from 0 percent to 40 percent. In contrast, stimulating the left hemisphere (and inhibiting the right hemisphere) yielded little or no improvement.

Many questions remain about the methods and results of these studies. At face value, they suggest that direct stimulation of the brain can facilitate insight, or can at least temporarily shift one's cognitive style from analytic to insightful. However, much more research is necessary to verify and understand these findings before it can be determined when this technique might be helpful and when it might have undesirable effects. For example, brain stimulation can sometimes enhance one ability while suppressing another.

For some people, the idea of enhancing insightfulness by running an electric current through their brains will seem scary, even creepy. Nevertheless, imagine the implications. In a few years, problem solvers may have the option of donning high-tech "thinking caps" to help them overcome impasses and achieve insights. This could be particularly helpful when an insight is needed in a stressful situation, such as when military commanders and intelligence analysts must quickly grasp the big picture and think of a creative strategy. Time will tell.

So what can you do right now to enhance your creative insight? As we've shown, insight training can improve your ability to solve problems analytically but is unlikely to produce the kind of spontaneous aha moments that can occur when you don't realize that you are stuck or when you don't recognize that you have a problem to solve. Brain stimulation is experimental and will not be available as a safe, reliable method for enhancing insightfulness for years, if ever.

There is a third approach. You can use the many research findings that we've discussed throughout this book to help you design an environment that will induce in you a brain state that is amenable to aha moments. You can do this just when you need to solve a specific problem or, without a particular problem in mind, you can change your lifestyle to generally maintain your openness to spontaneous realizations. Let's pull together what we know about how to promote this state, both generally and on demand.

MAKING IT HAPPEN

You are more likely to have creative insights and valid intuitions when your brain is in a general state characterized by remote associations, broad flexible attention, abstract thought, positive mood, a sense of psychological distance, and a promotion orientation.

Expansive surroundings will help you to induce the creative state. The sense of psychological distance conveyed by spaciousness not only broadens thought to include remote associations, it also weakens the prevention orientation resulting from a feeling of confinement. Even high ceilings have been shown to broaden attention. Small, windowless offices, low ceilings, and narrow corridors may reduce expenses, but if your goal is flexible, creative thought, then you get what you pay for.

Small-scale features of your surroundings can also broaden or narrow your attention. Objects with edges that merely appear sharp, even if they are on soft objects such as sofas, will constrict your mind by automatically eliciting in you a subtle, unconscious feeling of threat. (Keep that letter opener in the drawer!) More generally, your surroundings should not have striking objects or features that imply threat or that grab and focus your attention. Thus, the ideal environment for creative thinking is open, airy, soft, rounded, and calm.

Relaxing outdoor colors such as blue and green contribute to this state; "emergency" colors such as red suppress it. Darkness or dim lighting also works because they obscure visual detail and shift thought toward generality, abstraction, and broad attention. This principle likely applies to the other senses as well, so relatively quiet surroundings are preferable. But when silence would lead to excessive drowsiness or when some noise is unavoidable, ambient noise should be soft, diffuse, and resistant to focus, such as the sound of ocean waves or the simultaneous murmuring of many people. Undoubtedly, one of the reasons why so many creative people like to work in coffee shops is the murmuring and soft innocuous music found there. New Age music, another favorite of some creative types, often features relaxing nature sounds and a listening experience that is both expansive and introspective.

Though your creative environment should be perceptually soft and diffuse, it shouldn't remain unchanging. Static surroundings encourage static thinking. Don't be predictable. Don't get stuck in a rut. Enforced change and adaptation will destabilize entrenched thoughts through fixation forgetting to lubricate your mind for breakthrough ideas. To accomplish this, you should occasionally change everyday routines, such as where you go for coffee or the route you take to get to work. At home and in your workplace, rearrange your furniture or change the décor from time to time. Hold meetings in a variety of places. These kinds of changes may slightly decrease your day-to-day efficiency, but transformational insights can ultimately increase your productivity more than enslavement to habit.

Instead of surrounding yourself with people like you, interact with diverse individuals, including some (nonthreatening) nonconformists. Unusual people are not only out-of-the-box thinkers, but their mere presence primes a deviance mind-set that will enhance insightfulness in you and your more buttoned-down coworkers. This effect is magnified by the presence of communal areas where you and your colleagues must interact. Such water-cooler conversations are also incubation breaks that induce fixation forgetting and provide insight triggers.

Seek out symbols of others' creative achievements as well as reminders of your own past creative exploits. Aspects of foreign cultures such as art, food, and music from around the world are also helpful, especially if they bring to mind memories of your own efforts to assimilate or adapt to an unfamiliar culture.

The threat of a firm deadline will narrow your thinking and inhibit your insightfulness. Have the patience to wait for a breakthrough idea. When this is impractical, minimize mental constriction and the prevention orientation that it spawns by using soft target dates and a flexible schedule to establish a helpful, nonthreatening time frame. Rewards and punishments for meeting or missing deadlines, if needed at all, should be vague and mild so that they don't grab and squeeze attention. The rush of a sudden insight is usually reward enough.

Even if you don't have the luxury of sculpting your environment in these ways, a few simple lifestyle modifications can provide a creative boost. Periodically consider your larger goals and how to accomplish them; merely thinking about this will induce a promotion mind-set. Reserve time for long-range planning. Thinking about the distant future stimulates broad, creative thought. Cultivate a positive mood by thinking about people and things that bring you joy. To put a twist on Pasteur's famous saying, chance favors the happy mind.

When you are stuck on a problem, take a break to do or think about something very different. The initial failure to solve a problem can sensitize you to things that can trigger an insight, so, during your break, expose yourself to a variety of people and places or change your activity. Perhaps listen to music or go to a pleasing movie, art exhibit, or creative performance. Play a game or attend a sports event. Dance. Do yoga. Read. Just do *something* else. Insight triggers appear at the most unlikely times and places to reveal analogies and other hidden connections among your ideas and experiences.

The emergence of an insight is often immediately preceded by a period of reduced awareness of one's surroundings. Even if you don't have a specific problem to solve, alternate periods of externally focused attention with time spent in quiet, peaceful surroundings focused on your own thoughts. Internal focus will minimize distractions that can block an unconscious idea from popping into your mind. So take a walk, preferably in a peaceful environment; meditate; or go somewhere tranquil to sit quietly and introspect. If this isn't possible, simply close your eyes.

Ample sleep magnifies fixation forgetting and empowers memory consolidation to foster the discovery of hidden connections between ideas. There's no shame in taking a nap; sleep is creative mental work. Tell your boss that we said so.

Analytic thought is at its best during your peak time of day because this is when your brain's ability to inhibit irrelevant thoughts and maintain focus is optimal. Insightful thought is at its best when your powers of inhibition are weaker, because reduced focus opens your awareness to remote associations and seemingly irrelevant insight triggers that would otherwise be excluded when you are sharper. So if you're an early bird, you should try doing your creative work at night. If you're a night owl, try the morning.

When you are fixated on a personal problem that is dragging you down, it's especially helpful to use all of the strategies that we have described for broadening your thinking. Not only will this help you to find an insightful solution by reminding you of the big picture, but it will also disrupt the anti-insightful grip of rumination and depression. And do whatever it takes to reduce anxiety to avoid mental constriction.

Perhaps the single most important thing to remember is that your mental states can change, but they can't turn on a dime. It takes a while to sink into an insightful mind-set. This means that switching back and forth between insightful and analytic thought can waste time and prevent you from fully engaging either state. Try to schedule uninterrupted blocks of time for relaxed, freewheeling creative thought. Turn off your cellphone. Get rid of the clock. Let abstract ideas and vague impressions flow where they will. Be open to the patterns that emerge.

These strategies will help you broaden your mind and prepare it for new realizations. You can apply them informally and on an ad hoc basis, or you can be more systematic. One way to accomplish this is through self-experimentation. This has two key components. The first is to implement a single change. For example, you can alter a part of your regular routine to disrupt entrenched thought patterns, or you can take fifteen minutes each day to do long-range planning to prime psychological distance, or you can schedule an extra hour of sleep every night to promote incubation.

The next component involves measuring the effects of your change. To do this, you must have a measurable outcome. For example, if you're a writer, note how many pages of high-quality writing you are producing each day. If you are a student, note how many insightful comments you make in your classes.

An alternate strategy is to keep a diary of problems, solutions, and ideas. Start by making a list of problems that you want to solve, including both long-standing problems and new ones. As you systematically apply the strategies that we have described, keep a record of ideas and solutions that occur to you. Take special note of new problems that you might think of during the experiment itself, because realizing a problem that you were previously unaware of can itself be a creative act that should be documented.

It's important to test the effectiveness of each change that you implement. You shouldn't just assume that a change is helping. In a laboratory study, this would be done by comparing the behaviors of an experimental group of participants with those of a control group. For example, the experimental group might do some distant-future planning each day that the control group doesn't do. Differences in the creative outputs of the two groups would then be noted.

This standard laboratory technique doesn't work in self-experimentation because you have no control group. However, you can be your own control group by using an "A-B-A" strategy: Measure your creative output before making a change (condition A); measure it after implementing the change (condition B); and then go back to the beginning by removing the change and measuring your creative output again (condition A again). If the change helped, then there should be an increase in creative output during condition B compared with the first condition A. This improvement should disappear when the change is removed in the repeat of condition A.

The approach of compiling a list of changes to try and then testing them one by one is systematic and effective, but it's also a slow process. It could take a long time to isolate the changes that would be most helpful for you as an individual. A shortcut strategy is to implement several creativity-enhancement changes all at once or even integrate them into a single complex intervention. This may not reveal which specific changes were most helpful, but it can potentially produce the quickest benefits. Here's a real-life example of this kind of intervention.

HITTING A HOLE IN ONE

One spring night, Greg Swartz, director of innovation for golf equipment manufacturer Ping, sat on the back porch of his home outside Phoenix, Arizona. His goal was to come up with an improved golf club design. Swartz shut off the lights. He leaned his chair back and gazed up into the starry sky. Actually, he stared past the starry sky, in his words, "focusing into infinity." It was a quiet night. He put on some 1980s songs as background music, tunes with which he was familiar but to which he felt no particular emotional attachment. Soon, his thoughts began flitting about:

> *Golf clubs . . . People want to hit the ball longer and faster. . . . What are the physics of golf? . . . Trampolining . . . two springs . . . a system: golfer, ball, shaft, head . . . It's a system. . . . What is a system? In golf, the swing is part of a system that rotates around your wrist. So, if you tighten the arc in between the shoulders, the axis of rotation comes closer to your hands. So the system is moving faster if you shorten the axis of rotation. . . . Like a spinning figure skater, spinning faster when her arms are pulled in, tightening the axis of rotation . . . It's like the hammer throw in track and field, spinning faster for more force.*

His thoughts buzzed about until he remembered that there were constraints: the U.S. Golf Association rules. Swartz considered all these constraints: club speed, weight, and so forth. *But wait,* he thought. *This is a system.*

> *How fast can I deliver the club to the ball? The shaft is a spring. What is a good analogy? There has to be a good one. . . . Spring . . . fishing! Fishing rods and casting. Spring coils and releases. Casting with a little wrist flick. What about a much tighter rodlike deep-sea fishing? Can you cast? Not well. The shaft is so tight you can't get it to spring. Poor cast. Why are these things different? Weight on end versus springiness in rod. Those are constraints. If it's a light lure, it's the tip that makes it snap, but if it's heavy—like a golf club—then it's the piece near the end that matters. It's a projectile you're trying to launch. OK, the golf swing is also a system. Can I change the system closer to the hand? . . . Most of the constraints from USGA are on the clubhead, so the system can be changed elsewhere!*

That was, Swartz says, the aha moment he needed. From that moment, he knew where he was going. He just needed to figure out a specific way to implement his idea. He quickly conjured up two potential solutions. (For competitive reasons, he didn't reveal the specifics of his ideas to us.) There were some loose ends to tie up, but nothing problematic. You may be able to find his innovation at your local golf store by the time you read this book.

What's particularly interesting about this aha moment is that it wasn't a random occurrence. Rather, it was the result of a well-considered intervention to shift his brain into a state conducive to insight.

Coming up with creative solutions is Swartz's job as Ping's Director of Innovation. Before taking this position, Swartz had a track record of innovation in several industries, including aerospace engineering, software development, and musical-instrument manufacturing. He said that earlier in his career he came up with ideas somewhat haphazardly. He wasn't able to generate them on demand and had to patiently wait for them to emerge. Moreover, he didn't fully trust the validity of his ideas. So, when a new one occurred to him, he would put it aside for a few months before returning to re-evaluate it afresh. Then, if he still thought the idea had merit, he would pursue it.

Swartz said that during this period most of his insights were relatively minor and his progress slow and uneven. He therefore made a conscious effort to develop his personal technique for generating aha moments. For our purposes, the specifics of his method—the porch, the stars, the background music, and so forth—are not the central concern. Swartz could have selected other features to help him achieve the mental state he seeks. Rather, the general principles of insight enhancement that we've discussed throughout this book are key. To understand this, let's examine his technique in more detail.

Hermann von Helmholtz said that it was important to have "turned my problem over on all sides to such an extent that I had all its angles and complexities 'in my head.'" Greg Swartz described a similar process to familiarize himself with the USGA rules on club design and the factors that make a golf ball fly straight and far. This much is common sense. You've got to do your homework before you can innovate. Only after reviewing this information would Swartz try to establish a brain state conducive to insight.

He likes to do this at night, but not too late. "I'm not too ready to sleep. I make sure everything is set; get myself in a frame of mind. Might look at a few notes on club design, but I've immersed myself enough. It's all in my head. I need to be relaxed, but my mind alert. I'll sit out back and bring cold water [to prevent drowsiness]. My process is like what your body goes through before you fall asleep. You're no longer aware of some things—almost like sensory deprivation. It's the same thing that people do when they're really focused—watching a movie they're really into, not cognizant of their body, really focus on their thoughts. Me, I'm focused at will." (Swartz uses the term "focus" to refer to a state of absorption in his own thoughts.)

This is reminiscent of our experimental findings on preparation for insight. We found that just before viewing a problem that participants would eventually solve with insight, they disengaged from their surroundings and directed their attention inwardly on their own thoughts. In contrast, just before viewing a problem that they would eventually solve analytically, participants expectantly directed their attention outwardly on the screen on which the problem would be presented.

The environment Swartz has constructed helps to reinforce this state of inner-directed attention. The mild temperature and comfortable, slightly reclined chair allow him to relax and focus on his thoughts rather than on his surroundings. The background music doesn't grab his attention but does obscure potentially distracting sounds: "The music is like a mask; I don't hear it." He looks up at the sky but doesn't focus on the stars. He gazes past them into infinity and makes sure that there is nothing recognizable in his view. We asked if he knows the constellations. "No." If he did, might that interrupt his process? "It probably would."

This part of Swartz's technique exemplifies the principle that broad, diffuse attention facilitates creative insight. Gazing into infinity evokes psychological distance, which primes insightfulness. The darkness also helps because, in the absence of detailed sensory input, it nudges people to think in broader, more general terms.

Swartz prefers to furrow his brow, because he thinks that it may help if he feels slightly annoyed at the start of a session. This seems to contradict the research showing that a positive mood facilitates insight; however, rather than his initial irritation, it's probably the subsequent improvement in his mood brought about by his pleasant surroundings that really promotes his creative explorations.

Then, he starts thinking. *What am I looking for? Longer and faster.* But he doesn't pursue his thoughts in a deliberate, methodical fashion. Instead, he releases these thoughts to meander and collide. This enables him to view the problem from different perspectives and sets the stage for making connections among remotely associated ideas. Then he seeks analogies for the problem. But he's careful not to latch on to any one of them for too long. This prevents fixation on any ideas that will ultimately prove unproductive.

Only when he has his aha moment of clarity and certainty about the solution does he change tack. Even though he has already come up with his idea, he doesn't want to waste the state: "Once I'm in that mode, that focus, I want to take advantage of it. . . . Maybe I'll come up with something else. Once I get that, I drift around looking for other areas, other problems." Finally, after a few minutes, he goes back into his house and writes down each idea, making notes about what points are in need of elaboration.

Greg Swartz's personal method exemplifies a disciplined effort, based on years of self-experimentation, to induce a brain state that is receptive to insights. His specific technique—incorporating the porch, the stars, the music—is just one possible instantiation of some of the scientific principles that we have described throughout this book. You can tap these same principles to design your own personal technique to maximize your own creative potential. Just be sure to test its effectiveness on yourself so that you can optimize it.

Of course, life in the twenty-first century is busy, frenetic, and overstimulated. You may not have the flexibility to dedicate blocks of time to sink into an insight mind-set the way Greg Swartz does. If this describes your life, then the next best thing is to repurpose time that you already spend on routine, undemanding activities.

For example, John has found a way to carve out a creative space for himself whenever he rides on a train. Commuter trains are often noisy and crowded—a far cry from an insight-friendly environment. He isolates himself by wearing sunglasses, closing his eyes, and using noise-canceling headphones to block out the sound of the train (rather than to play music). On the noisiest, most crowded trains, he uses earplugs as well as noise-canceling headphones. To eliminate the constricting fear that he might miss his train station, he connects the headphones to his iPod and sets the device's alarm to gently alert him a few minutes before his stop. Then, in the relative darkness and quiet, with his cellphone turned off and no Internet access, he spends a few minutes inducing an insight mind-set by thinking about the distant future, about things that make him happy, and so forth. Once he feels that his mind is open and defocused, he thinks of a problem that he wants to solve or a goal for a new idea. Then he relaxes his thoughts further and allows the associations to flow. He's gotten many of his best ideas in this way.

What time can you repurpose for creativity? Any time in which you are engaged in no task, such as when you are waiting in line, or are doing a task that demands minimal attention, like housework or gardening, can be commandeered for creative work in this way. The classic example is the isolation of the shower. Meal times are another possibility. (After all, you have to eat.) Try taking your breakfast or lunch alone in a quiet, isolated environment. Start by priming yourself for a positive mood, psychological distance, and a promotion orientation, for example, by thinking of what long-range goals would make you happy. Then let the ideas flow while you unhurriedly nibble on your meal. Thinking about a problem while doing an easy task may not be particularly convenient. However, incubating a problem while doing a simple, unrelated task can help more than incubation during either a demanding task or no task.

These strategies and principles, if applied systematically, will enhance your insightfulness. But if you want to maximize it, then it's important to go a bit deeper.

ENEMIES OF THE STATE

When you perform a task, areas of your brain involved in accomplishing the task become more active compared with their resting levels. Which specific areas "light up" depend on the type of task: Visual areas in the back of the brain kick in during tasks that involve looking at things; motor areas in the frontal lobe are engaged during tasks that require physical movement; and so forth. This principle underlies many neuroimaging experiments.

A few years ago, neurologist Marcus Raichle and his collaborators began to look at the flip side of this: brain areas that might be more active when a person is *not* engaged in a task. What they found has far-reaching implications: a whole "default state" network of brain regions that are more active while a person is at rest. These areas become energized when your mind disengages from your surroundings to consider psychologically distant inner worlds: past events, future events, another person's thoughts, faraway places, and so forth—just the kinds of mental travel that prime you to think insightfully. To construct these novel internal worlds, you must direct your attention inwardly and expand your conceptual attention to include ideas that are remotely associated with the world around you. This is the state that the would-be Insightful seeks.

But mental travel through the multiverse of the default state confers benefits beyond the fact that you are more likely to have insights while visiting psychologically distant realms. We've seen that aha moments don't pop into existence from nothing. They are sudden, conscious products of unconscious processes that can take minutes—or years—to come to fruition. The catch is that they don't always come to fruition. Your intuition may tantalize you by hinting that a novel thought lies just out of reach of your conscious mind. But that's no guarantee that the idea will show itself. The challenge is how to smooth an insight's path to awareness. There's no point to your brain's having a novel idea if you don't know what that idea is.

The default state helps propel your nascent ideas into consciousness even when you aren't in that state. Helmholtz's nature walks, Descartes's late mornings in bed, Jerry Swartz's playtime with his children, and Judah Folkman's contemplation at his synagogue all culminated in moments of insight. But these types of activities are also powerful incubators for ideas that reveal themselves at other times. They also broaden thought and impart a creative momentum that persists even after one reengages with the world.

The dilemma is that the default state is fragile. It doesn't take much for the outside world to puncture your introspective reverie, quickly deflating your default-state network and thrusting to the forefront other brain networks that immediately set to work dealing with the demands of the moment. This is how it has to be. If humans weren't quick to respond to all the opportunities and threats around us, we wouldn't have survived.

However, modern life has put unforeseen strains on our innate responsiveness. It demands nearly constant attention to the environment. There is continuous pressure to produce, especially during economically uncertain times. Meanwhile, many people strive to be attentive parents, spouses, partners, relatives, or friends. Leisure time is hard to come by, and it's difficult to resist expending what there is of it on the alluring social media platforms, games, videos, gadgets, podcasts, and television channels that are constantly clamoring for our attention. Our thoughts can be interrupted by a cellphone or text message anywhere and anytime. And when the cellphone isn't ringing and there aren't any new text or email messages, we think about when the next call or message will arrive. The twenty-first century is all about staying connected, staying on top of things, not missing anything. This is exciting and often productive, but we need to be aware of all of its effects on us.

The truth is, we live in an environment on steroids, far removed from the more relaxed pace of life just one generation ago. We are slaves to the present moment, with little opportunity for quiet introspection and mental travel. The inner world of the default-state network hardly has a chance. As a society, we are trading creativity for a narrow type of efficiency.

We haven't been dragged kicking and screaming into this obsession with the here and now. As a species, it seems to be our preference. When we are required to think about other worlds, we prefer to think about the recent past, the near future, people like us, and scenarios closely related to our current reality. The Insightfuls among us may find it somewhat more natural to think about distant worlds, but even their mental travel tends not to stray very far. And for everyone, the inner world is a fragile thing, easily overwhelmed by the glare and allure of the outer world. It is easily crushed and hard to revive. Consider the implications.

For all the concern over the "innovation gap" and declining creativity, a crucial cause is overlooked: our addiction to constant stimulation and a relentless drive for (noncreative) productivity amplified by technology and global competition. Hyper-connectedness, hyper-competitiveness, hyper-availability, hyper-instability. All 24/7. All enemies of the (default brain) state.

Of course, the solution isn't to go through life as a sensory-deprived hermit. Modern life can stimulate, as well as sap, one's insightfulness. We've seen that alternating between the inner and outer worlds is the best way to enhance your creativity. The impediment to achieving this balance is that the outer world keeps its thumb on the scale. The only way to compensate is by giving extra weight to the inner world and by periodically banishing its enemies. This can be done—we have seen many examples of Insightfuls who have shown us how. Follow their example, and you will be able to realize the power of the state.

THE PATH FORWARD

We began this book with the story of the aha moment that started Helen Keller on her path to enlightenment and glory. Later, she had another insight that continued her transformation into what Winston Churchill called "the greatest woman of our age."

One day, Anne Sullivan tried to teach young Helen to string beads according to a pattern. Helen, not fully grasping the pattern, made a number of mistakes, which Anne showed to her. One error in particular perplexed Helen. Anne, sensing Helen's concentration, touched the girl's forehead and then spelled on her hand the word "Think." Helen later wrote, "In a flash I knew that the word was the name of the process that was going on in my head. This was my first conscious perception of an abstract idea."

Such insights empowered Helen to explore and colonize a new and ever-expanding realm of ideas. As Mark Twain wrote, "Napoleon tried to conquer the world by physical force and failed. Helen tried to conquer the world by the power of mind—and succeeded!" She and her fellow Insightfuls have shown that the power of this state is not restricted to the privileged few. It's available to everyone, to you.

Create the state. It's the empire of the future.

What will your insights be?

ACKNOWLEDGMENTS

With gratitude, we would like to acknowledge the help and support we have received from many sources. Thanks go to the National Institutes of Health, which supported both of us early in our careers, and to the National Science Foundation and the John Templeton Foundation, which currently fund John's and Mark's labs, respectively. We also thank the students and collaborators who have worked with us on insight research over the years, including Julia Anderson, the late Stella Arambel-Liu, Edward Bowden, Brian Erickson, Jessica Fleck, Deborah L. Green, Richard Greenblatt, Jason Haberman, Roy Hamilton, Monica Truelove-Hill, Todd Parrish, Lisa Payne, Paul Reber, David Rosen, Roderick W. Smith, Jason van Steenburgh, Jennifer Frymiare Stevenson, Karuna Subramaniam, and Ezra Wegbreit. We are grateful to Demetrios Voreades of Source Signal Imaging, Inc., to John Kimura of Sensorium, Inc., and to Teunis van Beelen for assistance with technical aspects of EEG recording and analysis.

We are appreciative to the news media for bringing our work to the public's attention. In particular, *The Times* of London, NPR, *U.S. News & World Report, The Philadelphia Inquirer,* and *O, The Oprah*

Magazine produced some of the first reports of our work. *The New Yorker, The Wall Street Journal, The New York Times,* and *Scientific American Mind* provided more recent coverage. Most recently, the BBC Television series *Horizon* produced an episode on insight featuring some of our research.

We also thank Suzanne Gluck, our literary agent; Will Murphy, our editor at Random House; Susan Kamil, Random House publisher; and Ben Steinberg, Samuel Nicholson, Mika Kasuga, Janet Wygal, Michelle Daniel, Joseph Perez, Tom Hallman, and Casey Hampton of Random House for making this book happen. We are honored and delighted to work with them.

JOHN'S ACKNOWLEDGMENTS

I would like to thank friends, collaborators, mentors, colleagues, students, and supporters who have helped over the years, including Andy Burnett, Cindi Burnett, William Gehring, Jack Gelfand, Murray Grossman, Thomas Hewett, Keith Hokyoak, Phillip Holcomb, Marcia K. Johnson, Eric Kandel, Michael Kane, Scott Barry Kaufman, Youngmoo Kim, Howard Lavine, Michael Lowe, Laura McCloskey, David E. Meyer, Dan Mirman, Banu Onaral, Allen Osman, Robi Polikar, Suri Rajneesh, the late Seth Roberts, Dan Ruchkin, Aleister Saunders, Dan Schacter, Jonathan Schooler, Myrna Shure, Chris Sims, Saul Sternberg, Cynthia Thomsen, Michael Vogeley, Dan Willingham, and the late Steve Yantis. I would also like to thank Princeton University's psychology department, especially Sam Glucksberg, Phillip Johnson-Laird, Daniel Osherson, department chair Deborah Prentice, and Nicholas Turk-Browne for hosting me during a sabbatical that allowed me to do much of the work on this book. Thanks also go to past Drexel psychology department head Kirk Heilbrun, Dean Donna Murasko, Provost Mark Greenberg, and President John Fry for supporting my research. Special thanks go to my current department head,

James Herbert, for his wonderful encouragement and support. I would also like to express gratitude to Constantine Papadakis, the late president of Drexel University, and to Eliana Papadakis, for their support and inspiration. Thanks also to Joseph and Maria Armenti, Nicholas and Alexis Souleles, Steven and Maureen Stavropoulos, and Andrew and Catherine Walsh for their kind encouragement.

Love and thanks go to my wife, Yvette Kounios, and to our children, Bill and Daphne, for their encouragement, love, support, and enthusiasm during the writing of this book. I am also deeply grateful to Yvette for her exceptional editorial and creative input. Her expertise is evident throughout the book. I would also like to thank my father, Vasilios Kounios; my late mother, Stavroula Kounios; my mother-in-law, Era Stavropoulos; and my father-in-law, Vasilios Stavropoulos, for their unlimited love and support.

MARK'S ACKNOWLEDGMENTS

I would like to thank all the mentors and colleagues who taught and inspired me along the way. Among those who helped train me were Rhonda Friedman, Morton Ann Gernsbacher, Jordan Grafman, Doug Hintzman, Mike Posner, Art Wingfield, and Edgar Zurif. Many colleagues have helped in various ways, some specifically with research. Edward Bowden deserves special thanks. Not only did he help in the early days of my foray into insight research (and on and off since), but he was a good friend and all-around helpful presence during the times he worked in my lab. Thanks also go to Kalina Christoff, Melissa Ellamil, Sohee Park, Jonathan Schooler, and other researchers in the field who have inspired or collaborated on investigations of insight and creativity. Northwestern University's department of psychology has been a welcoming home for the past decade or so. My Brain Behavior and Cognition Program colleagues have been very supportive (of work and life), especially Marcia Grabowecky,

Ken Paller, Paul Reber, and Satoru Suzuki. Within that program, I've had the pleasure of helping train some bright and enthusiastic students who have contributed to my work and to my thinking: Azurii Collier, Lisa Hechtmann, Heather Mirous, Chivon Powers, Karuna Subramaniam, Ezra Wegbreit, and Darya Zabelina. Numerous other lab alumni have helped with the research, but Jason Haberman was a particular dynamo in helping with the early days of neuroimaging work, and I expect great things of his career.

Finally, with all my heart, I thank my kids, Oona and Elliot, for being who they are. What a wonderful gift.

NOTES

CHAPTER 1: NEW LIGHT, NEW SIGHT

3 The opening quote by Helmholtz is derived from Leo Koenigsberger's *Hermann von Helmholtz,* translated by Frances A. Welby (Oxford: Clarendon, 1906), 208. archive.org/stream/hermannvonhelmho00koenrich/hermannvon helmho00koenrich_djvu.txt

3 Helen Keller told the story of her insight about words in her autobiography, *The Story of My Life* (Garden City, N.Y.: Doubleday, 1921), 20–24. Anne Sullivan's description is derived from one of her letters published in the same book (316).

5 The insights of Sir Isaac Newton and the Buddha have been documented innumerable times. Sir Paul McCartney has described on CNN's *Larry King Live* how the melody of the song "Yesterday" came to him in a dream: www .youtube.com/watch?v=062Mz-y3hkU.

5 The scientific approach to studying the mind is explained in any good textbook about cognitive psychology or cognitive neuroscience. A fine example is E. B. Goldstein, *Cognitive Psychology: Connecting Mind, Research and Everyday Experience,* 3rd ed. (San Francisco: Cengage Learning, 2011).

A Matter of Interpretation

8 For background about Gestalt psychology and its origins, see *Wikipedia,* s.v. "Gestalt psychology," last modified July 6, 2014, en.wikipedia.org/wiki/

Gestalt_psychology/. A summary of early Gestalt research on insight can be found in R. E. Mayer, "The Search for Insight: Grappling with Gestalt Psychology's Unanswered Questions," in *The Nature of Insight,* ed. R. J. Sternberg and J. E. Davidson (Cambridge, Mass.: MIT Press, 1995), pp. 3–32.

8 Information about the Wright brothers' propeller design can be found here at en.wikipedia.org/wiki/Wright_brothers#cite_note-47.

Insight Is Creative

9 For a recent discussion of definitions of insight, see J. Kounios and M. Beeman, "The Cognitive Neuroscience of Insight," *Annual Review of Psychology* 65 (2014): 71–93.

9 For a discussion of definitions of creativity, see J. C. Kaufman, *Creativity 101* (New York: Springer Publishing, 2009).

10 Discussion of insight as a sudden process can be found in R. W. Smith and J. Kounios, "Sudden Insight: All-or-None Processing Revealed by Speed-Accuracy Decomposition," *Journal of Experimental Psychology: Learning, Memory, and Cognition* 22 (1996): 1443–62.

Everyone, Everywhere

11 Clarence Birdseye's sudden insight into how to freeze food without ruining it is described on page 26 of R. Platt, *Eureka! Great Inventions and How They Happened* (Boston: Kingfisher, 2003). Philo Farnsworth's insight about television is described in chapter 1 of P. Schatzkin, *The Boy Who Invented Television* (Silver Spring, Md.: TeamCom, 2002).

11 In labeling creativity as an essentially human trait, we did not discuss research on insight in animals. Some animals may have a rudimentary form of creative insight, just as some animals have the capability of acquiring a rudimentary form of language. But the difference in the scale of animal and human creativity is vast, just as it is for language. For a balanced discussion, see S. J. Shettleworth, "Do Animals Have Insight, and What Is Insight Anyway?," *Canadian Journal of Experimental Psychology* 66 (2012): 217–26.

12 The quotations from D. T. Suzuki about satori can be found in D. T. Suzuki, *An Introduction to Zen Buddhism* (New York: Grove Press, 1964).

12 The results of the Pew survey on religion can be found at the Pew Forum: pewforum.org/docs/?DocID=490#6. A discussion of sudden religious conversion can be found in "The Moment of Truth," *The Economist,* July 24, 2008.

Metamorphosis

13 Carey's research on sudden changes during psychotherapy is described in T. A. Carey et al., "Psychological Change from the Inside Looking Out: A Qualitative Investigation," *Counselling and Psychotherapy Research* 7 (2007): 178–87.

13 The quote from Oliver Wendell Holmes, Sr., is derived from *The Professor at the Breakfast-Table* by Oliver Wendell Holmes (page 235): archive.org/stream/professoratbrea12holmgoog#page/n233/mode/2up/search/experience.

The Information Vortex

13 The IBM survey of CEOs is described in IBM Global Business Services, *Capitalizing on Complexity: Insights from the Global Chief Executive Officer Study* (2010).

Eurekanomics

14 Thomas Friedman's notion of the "flat world" is described in Friedman, *The World Is Flat: A Brief History of the Twenty-First Century* (New York: Macmillan, 2006). Judy Estrin, among others, has discussed the notion that the United States is experiencing an "innovation gap." See Claire Cain Miller, "Silicon Valley Entrepreneur Warns of U.S. Innovation Slowdown," *The New York Times,* September 1, 2008. For a discussion of innovation policy, see Steve Lohr, "Can Governments Till the Fields of Innovation?," *The New York Times,* June 20, 2009. Also see the website of the Institute for Large Scale Innovation, www.largescaleinnovation.com.

How Smart Are We?

16 The Flynn effect is summarized in *Wikipedia,* s.v. "Flynn Effect," last modified July 27, 2014, en.wikipedia.org/wiki/Flynn_effect.

16 The decline in U.S. creativity test scores is reported in K. H. Kim, "The Creativity Crisis: The Decrease in Creative Thinking Scores on the Torrance Tests of Creative Thinking," *Creativity Research Journal* 23 (2011): 285–95.

CHAPTER 2: INSIGHT ILLUSTRATED

Stepwise

17 A version of the stage model of insight was first explicitly described in G. Wallas, *The Art of Thought* (London: J. Cape, 1926).

Vantage Point *and* At That Moment

18 Robert Cooke's biography of Judah Folkman describes his insights: *Dr. Folkman's War: Angiogenesis and the Struggle to Defeat Cancer* (New York: Random House, 2001). Y. Cao and R. Langer recently summarized Folkman's contributions in a scientific article: "A Review of Judah Folkman's Remarkable Achievements in Biomedicine," *Proceedings of the National Academy of Sciences, of the United States of America* 106 (2008): 13203–5. PBS has produced two reports about Judah Folkman and his research. These can be viewed at the PBS.org website, along with a wealth of supplementary material.

The Flip Side

22 Information about the Mann Gulch fire can be found in R. C. Rothermel, *Mann Gulch Fire: A Race That Couldn't Be Won,* U.S. Department of Agriculture Forest Service Intermountain Research Station General Technical Report INT-299, 1993. Additional background can be found on the website of the Dodge Family Association: www.dodgefamily.org/Biographies/R/Robert_Wagner_Dodge.shtml.

By Way of Analogy

23 Andrew Stanton's story and quotes are derived from his July 10, 2008, NPR *Fresh Air* interview: www.npr.org/templates/story/story.php?storyId=92400669.

Self-Evident Truth *and* Certainly Joyful

26 The quotes from Barbara McClintock appear in E. F. Keller, *A Feeling for the Organism: The Life and Work of Barbara McClintock* (New York: W. H. Freeman, 1983).

27 Information about Archimedes can be found at Drexel University's Archimedes website: www.cs.drexel.edu/~crorres/Archimedes/contents.html.

More, Please

29 Carl Sagan's quote can be found in chapter 2 of Sagan, *Broca's Brain: Reflections on the Romance of Science* (New York: Random House, 1974).

29 The English translation of the quote from Hermann von Helmholtz appeared in R. S. Woodworth, *Experimental Psychology* (New York: Henry Holt, 1938), 818.

CHAPTER 3: THE BOX

31 The origin of the phrase "Columbus's egg" is discussed in *Wikipedia,* s.v. "Egg of Columbus," last modified May 14, 2014, en.wikipedia.org/wiki/Columbus_Egg. The story may be apocryphal, or the egg trick may have been performed by the Italian architect Filippo Brunelleschi.

32 Figure 3.1 (Columbus standing an egg on its end) was taken from J. Trusler, *The Works of William Hogarth: In a Series of Engravings with Descriptions, and a Comment on Their Moral Tendency* (London: Jones, 1833), en.wikipedia.org/wiki/File:Columbus_egg.jpg. Figure 3.2 (Christopher Columbus's Egg Puzzle) can be found in S. Loyd, *Cyclopedia of Puzzles:* en.wikibooks.org/wiki/Creativity_-_An_Overview/Thinking_outside_the_box#mediaviewer/File:Eggpuzzle.jpg.

33 An early classic study of the Nine-Dot Problem is described in N. R. F. Maier, "Reasoning in Humans: I. On Direction," *Journal of Comparative Psychology* 10 (1930): 115–43. An example of a recent cognitive psychology study of the Nine-Dot Problem is described in T. C. Kershaw and S. Ohlsson, "Multiple Causes of Difficulty in Insight: The Case of the Nine-Dot Problem," *Journal of Experimental Psychology Learning Memory and Cognition* 30 (2004): 3–13. Here is another solution to the Nine-Dot Problem. This one uses three, rather than four, lines:

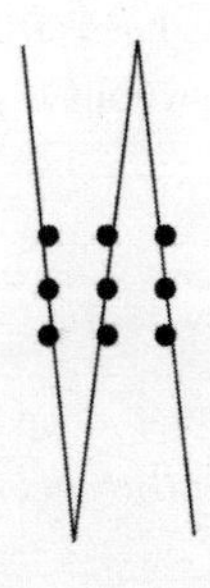

35 The classic study of the Two-String Problem was conducted by N. R. F. Maier, "Reasoning in Humans II: The Solution of a Problem and Its Appearance in

Consciousness," *Journal of Comparative Psychology* 12 (1931): 181–94. H. G. Birch and H. S. Rabinowitz's sequel to the original Maier study of the Two-String Problem is described in Birch and Rabinowitz, "The Negative Effect of Previous Experience on Productive Thinking," *Journal of Experimental Psychology* 41 (1951): 121–25.

In fact, even words can be a constricting box. Jonathan Schooler and collaborators showed that thinking out loud while trying to solve an insight problem can degrade one's ability to solve it. This phenomenon is called "verbal overshadowing" and seems to occur in situations in which solving the problem relies on nonverbal thought. This study is described in J. W. Schooler and T. Y. Engstler-Schooler, "Verbal Overshadowing of Visual Memories: Some Things Are Better Left Unsaid," *Cognitive Psychology* 22 (1990): 36–71.

What the Cortex Does

38 A primer on neuroscience can be downloaded for free from the website of the Society for Neuroscience: www.brainfacts.org/about-neuroscience/brain-facts-book/. The picture of an EEG brainwave in figure 3.6 is derived from en.wikipedia.org/wiki/File:Eeg_raw.svg.

Built for Surprise

41 The discovery of the N400 response by Marta Kutas and Steven Hillyard is described in Kutas and Hillyard, "Reading Senseless Sentences: Brain Potentials Reflect Semantic Incongruity," *Science* 207 (1980): 203–5. Figure 3.7, showing an N400, is based on the following study by Phillip Holcomb and John Kounios et al., "Dual Coding, Context Availability and Concreteness Effects in Sentence Comprehension: An Electrophysiological Investigation," *Journal of Experimental Psychology: Learning, Memory, and Cognition* 25 (1999): 721–42. Another important surprise EEG response is the P300: S. Sutton et al., "Evoked-Potential Correlates of Stimulus Uncertainty," *Science* 150 (1965): 1187–88. Two recent articles that discuss the view that a central function of the brain is to predict events are K. D. Federmeier, "Thinking Ahead: The Role and Roots of Prediction in Language Comprehension," *Psychophysiology* 44 (2007): 491–505; and J. Ghajar and R. B. Ivry, "The Predictive Brain State: Asynchrony in Disorders of Attention?," *The Neuroscientist* 15 (2009): 232–42.

Child's Play

44 The study showing that younger children can solve some types of problems more quickly and accurately than older children is described in T. P. German and M. A. Defeyter, "Immunity to Functional Fixedness in Young Children," *Psychonomic Bulletin and Review* 7 (2000): 707–12. The diagram of the brain shown in figure 3.9 is reproduced from commons.wikimedia.org/wiki/File: Brain_headBorder.jpg.

46 The study of problem solving in patients with frontal lobe damage is described in C. Reverberi et al., "Better Without (Lateral) Frontal Cortex? Insight Problems Solved by Frontal Patients," *Brain* 128 (2005): 2882–90.

Think Smart, Not More

48 The classic study of the cognitive foundations of expertise in chess was written by A. D. de Groot, *Thought and Choice in Chess,* 2nd ed. (The Hague: Mouton Publishers, 1965). A recent article discusses more recent perspectives and misquotations of de Groot's work: M. Bilalic, P. McLeod, and F. Gobet, "Expert and 'Novice' Problem Solving Strategies in Chess: Sixty Years of Citing de Groot (1946)," *Thinking and Reasoning* 14 (2008): 395–408. Regarding Bent Larsen's approach to playing chess, see query.nytimes.com/gst/fullpage.html ?res=9403E0DD1638F931A2575AC0A9669D8B63.

Quick Think

51 The information about, and quotes from, Captain Chesley Sullenberger are derived from www.airspacemag.com/flight-today/Sullys-Tale.html/; www .cbsnews.com/news/flight-1549-a-routine-takeoff-turns-ugly/; and *Wikipedia,* s.v. "Chesley Sullenberger," last modified June 26, 2014, en.wikipedia.org/ wiki/Sullenberger.

CHAPTER 4: ALL OF A SUDDEN . . .

58 The idea that creativity does not differ from "ordinary" thought is discussed in R. W. Weisberg, *Creativity: Understanding Innovation in Problem Solving, Science, Invention, and the Arts* (Hoboken, N.J.: John Wiley and Sons, 2006).

58 The anagram study by R. W. Smith and J. Kounios is described in "Sudden Insight: All-or-None Processing Revealed by Speed-Accuracy Decomposi-

tion," *Journal of Experimental Psychology: Learning, Memory, and Cognition* 22 (1996): 1443–62. Below is some additional background about this study.

To discover whether insights are sudden, we had to trace ideas backward in time to discover how they came into existence. We gave people a series of specially designed insight-like anagrams to solve. Each problem was a group of letters (for example, P-E-L-A-P) presented on a computer screen. Half of these letter groups could be rearranged to form real words (P-E-L-A-P = "APPLE"); the other half couldn't (P-E-L-E-P). For each puzzle, a participant's task was to quickly press one button if the letters could make a word or another button if they couldn't.

The critical part was that participants heard a beep after some of these letter groups appeared. This directed them to make their best guess immediately by pressing a button. We timed these beeps to force people to guess a fraction of a second before they would normally have solved the problem if they had been given more time. We wanted to see whether these slightly faster guesses showed better-than-chance accuracy. If the solution comes all at once as a sudden insight, then just before a person has the insight, she should have no information at all about the solution. (For example, on election night, you would have no idea who the winner is even an instant before you turned on the TV.) But if a person were accumulating information leading up to the complete solution, then a demand for a quick guess wouldn't leave her clueless. She would have some idea about the likely answer.

This experiment may sound simple, but it took many months to complete. All the while, we were nervously waiting for the results. We had already used this approach with several memory and language tasks, and all of these experiments showed that people had partial knowledge before achieving the solution. This made us think that our experiment would probably yield no evidence that insight is fundamentally different from analysis.

This experiment seemed like a long shot, but we thought that it was also our best shot. We figured that if people didn't solve these anagrams suddenly, then there was probably no such thing as sudden solving. If insight really were an illusion, then so be it.

Rod Smith ran the experiment and fed the raw data into a computer program that did some complex analyses. I anxiously waited for the results.

Now, you're probably expecting to hear that people solved these problems by sudden insight and that there was no evidence of gradual problem solving. That's not what we found. The numbers showed that people solved the problems gradually. There was no evidence of an aha moment. So much for insight?

But something didn't seem right. We then started looking at the results in much more detail, and what we saw perplexed us. It was as if our participants could guess the solutions to some of these problems with superhuman speed, so quickly that it began to look like they could guess some of the answers before the problems were even presented! Our inside joke soon became that we had become parapsychology researchers and had proven the existence of precognition. But once we stopped laughing, we were faced with a dilemma: What went wrong? We racked our brains to think of, and check, every possible error. The computer program that ran the experiment worked properly. The program that analyzed the data did what it was supposed to do. In fact, everything was working perfectly.

More jokes. Perhaps we would end up as fortune-tellers or nightclub magicians.

But once the laughing subsided, we were left with the same problem and the same anxiety. *What could have gone wrong?* This experiment was a part of Rod's doctoral dissertation. He had spent many months collecting and analyzing the data. What were we going to do now?

Then Rod had an idea. There was one thing that we hadn't checked. The commercial computer program that we used to control each experimental session "shuffled the deck" to present the puzzles to each participant in a random order. What if the program hadn't done a good job of shuffling the deck? Like expert card players, perhaps our participants had picked up on a subtle pattern that gave them a clue about whether some of the problems in the sequence had or didn't have a solution. (Remember, they only had to press a button to indicate whether there was a solution, not what the solution was.) If so, then the "advance warning" that this pattern afforded would be all they needed to give quick, reasonably accurate guesses when they were interrupted by a beep.

This didn't seem likely, but we had nothing else to go on, so we examined the sequences of problems to look for any patterns that our participants could have used as a clue. We found nothing. This was another big disappointment.

One possibility was left. Maybe there really was a pattern to the order of the problems—one that we couldn't see. At first this seemed unlikely: Why would our participants be able to pick up on a pattern that we couldn't see?

A few years before our study, cognitive psychologists began uncovering hard evidence for unconscious intuitive thought (Chapter 10). One example of this is the ability to detect and learn complex patterns without ever becoming consciously aware of them. These patterns, though unconscious, could still influence people's behavior. The participants in our experiment were tested over several sessions. Perhaps, with all this practice, they had picked up on *something* that gave them an advantage—something they weren't aware of.

We gave a lot of thought about how we could test this idea. However, it's hard to find something when you don't know what you're looking for. After much discussion, we settled on a plan. We feared that it could be a big waste of time, but we couldn't think of another way to proceed. So we went for it.

Rod obtained another computer program—a more sophisticated one—that generates random numbers. He used the new program, instead of the commercial experiment-running software, to shuffle the order of the puzzles. If the original shuffling had been the cause of our strange results, then a new batch of participants wouldn't be able to pick up on any patterns because there wouldn't be any patterns.

Then he ran the whole experiment over again. This took an extra eight months.

When Rod analyzed the new data, he found, to our relief, that our participants showed no evidence of gradual problem solving. He went on to repeat the experiment with a small change. Same results.

So, the original participants from the first experiment were not psychic. They had been picking up on a subtle pattern in the order of the problems, a pattern that provided them with clues about which anagrams did have a solution and which ones didn't. This taught us, up close and personal, the valuable secondary lesson that people can process information in a sophisticated way outside of awareness.

CHAPTER 5: OUTSIDE THE BOX, INSIDE THE BRAIN

63 Jerry Weintraub has described his Elvis insight in many interviews (for example, his April 20, 2010, interview with Tavis Smiley: http://www.pbs.org/wnet/tavissmiley/interviews/film-producer-jerry-weintraub/#top) and in his memoir: J. Weintraub and R. Cohen, *When I Stop Talking, You'll Know I'm Dead: Useful Stories from a Persuasive Man* (New York: Twelve, 2010).

65 Andrew Cohen's aha moment is described in www.independent.co.uk/arts-entertainment/music/features/mayer-hawthorne—haircuts-worked-it-all-out-1828332.html.

65 The Taylor Swift quotation is derived from the article "Taylor Swift's Telltale Heart," by Nacny Jo Sales, which appeared in the April 2013 edition of *Vanity Fair*.

"Problematic" Thinking

67 Remote associates problems were originally developed by Sanford Mednick: S. A. Mednick, "The Associative Basis of the Creative Process," *Psychological Review* 69 (1962): 220–32. The specific type of remote associates problems that we have used in our studies were compiled by Edward Bowden and Mark Beeman. See E. M. Bowden and M. Jung-Beeman, "Normative Data for 144 Compound Remote Associate Problems," *Behavior Research Methods, Instruments, and Computers* 35 (2003b), 634–39.

Facing Down the Clock

68 The findings about the errors of omission in high-insight thinkers and errors of commission in high-analytic thinkers are published in J. Kounios et al., "The Origins of Insight in Resting-State Brain Activity," *Neuropsychologia* 46 (2008): 281–91. We have obtained this same pattern of results in other studies.

When and Where

69 Figure 5.1 showing an fMRI scanner is reproduced from: picasaweb.google.com/lh/photo/4hL0Td00nq6kL73rJsiMAA. Mark and John's first neuroimaging study of insight is described in M. Jung-Beeman et al., "Neural Activity When People Solve Verbal Problems with Insight," *PLoS Biology* 2 (2004): 500–510, www.plosbiology.org/article/info%3Adoi%2F10.1371%2Fjournal.pbio.0020097. For a recent review of work on the cognitive neuroscience of insight, see J. Kounios and M. Beeman, "The Cognitive Neuroscience of Insight," *Annual Review of Psychology* 65 (2014): 71–93.

71 Figure 5.2 is adapted from M. Jung-Beeman et al., "Neural Activity Observed When People Solve Verbal Problems with Insight," *PLoS Biology* 2 (2004): 500–510.

Additional Thoughts About the Study

71 There is a widespread misunderstanding of how science works. Many people think that researchers propose a theory and then try to prove it. Sometimes it's done that way, but when it is, it's often not the most rigorous research. It's easy to find evidence to support a theory—scientific or otherwise—especially if you're selective about where you look. But supporting evidence should never be the sole factor that sways you to accept a particular theory. Instead, the best

way to validate an idea is by trying to disprove it. Throw everything at it. Give it every chance to fail. If the idea shows weaknesses, then make adjustments to it. If these adjustments don't help, then scrap the idea and start over. But if your idea holds up to all the criticisms that can be leveled at it, then you can start to feel confident that your idea has merit. Even then, you should stay vigilant for newly revealed weaknesses. (This approach can also be profitably applied outside of science.)

But we still ran our idea through the gauntlet to see if it would survive. Is there an alternate explanation for this right temporal lobe activity? Is it really central to insight? Perhaps it really reflects surprise at the solution, or pleasure at solving a difficult problem, or maybe it's just a part of the act of pressing the button to communicate that the problem has been solved.

These alternate explanations didn't hold water. For example, the gamma-wave burst can't reflect surprise or another emotion, because you can be surprised at something only after it happens; the gamma-wave burst occurred at the moment of solution by insight, not after. And it can't reflect processes involved in commanding fingers to press buttons. Neuroscientists have learned a great deal about the brain regions involved in planning and executing physical movements—this isn't one of them. So the timing and location of this brain activity argues strongly against alternate interpretations.

Another potential criticism of our conclusion that the gamma-wave burst is the neural manifestation of the moment of insight is based on a false observation. In a graph in our original 2004 *PLoS Biology* paper, it appears that the gamma-wave burst begins weakly and then grows over a short interval of time. This observation is the basis for the argument that this can't be an insight if it gets stronger. There are several reasons why this point is invalid. First, these results were computed with a mathematical technique called "wavelets." This method for computing the strength of oscillatory activity leaves a little bit of fuzziness around the edges. So when the gamma-wave burst appears to gain in strength, that is a by-product of the math rather than the brain. Second, the 2004 paper didn't actually find any evidence that the gamma-wave burst got significantly stronger over time, even if it tends to look that way in the graph. Third, and most important, even if the gamma-wave burst did get stronger over time, this would not be an argument against the view that this is the neural manifestation of insight. When you have an idea, it makes sense that it would take a fraction of a second for you to completely focus your attention on it.

CHAPTER 6: THE BEST OF BOTH WORLDS

72 D.B. and other patients with right-hemisphere damage are described in M. Jung-Beeman, "Semantic Processing in the Right Hemisphere May Contribute to Drawing Inferences," *Brain and Language* 44 (1993): 80–120.

72 Even in left-handed individuals, the left hemisphere is typically dominant for language. Estimates of left lateralization of language in left-handers range from 65 to 90 percent. These estimates are based on aphasia patients; procedures in which words are presented to either the left or right ear (directing most input to the opposite hemisphere) or to the left or right visual hemifield (directing input only to the opposite hemisphere); and a presurgical procedure that anesthetizes one hemisphere at a time.

73 Hiram Brownell and colleagues study a number of these phenomena in patients with right-hemisphere damage, such as those in the following pair of papers: H. H. Brownell et al., "Surprise but Not Coherence: Sensitivity to Verbal Humor in Right Hemisphere Patients," *Brain and Language* 18 (1983): 20–27; H. H. Brownell et al., "Inference Deficits in Right Brain-Damaged Patients," *Brain and Language* 29 (1986): 310–21.

What's the Difference?

76 Christine Chiarello has published many papers on lateralized semantic priming, including "Lateralization of Lexical Processes in the Normal Brain: A Review of Visual Half-Field Research," in *Contemporary Reviews in Neuropsychology,* ed. H. A. Whitaker (New York: Springer-Verlag, 1988), 36–76; and Chiarello et al., "Semantic and Associative Priming in the Cerebral Hemispheres: Some Words Do, Some Words Don't . . . Sometimes, Some Places," *Brain and Language* 38 (1990), 75–104.

Stepping Up

78 Children who recovered language function following left hemispherectomy to treat Rasmussen's syndrome are described in numerous published papers, including D. Boatman et al., "Language Recovery After Left Hemispherectomy in Children with Late-Onset Seizures," *Annals of Neurology* 46 (1999): 579–86; A. E. Telfeian et al., "Recovery of Language After Left Hemispherectomy in a Sixteen-Year-Old Girl with Late-Onset Seizures," *Pediatric Neurosurgery* 37 (2002): 19–21; and F. Vargha-Khadem et al., "Onset of Speech After Left Hemispherectomy in a Nine-Year-Old Boy," *Brain* 120 (1997): 159–82.

78 Many neuroimaging studies have found stronger left- than right-hemisphere activity. In fact, many reports have found only left-hemisphere activity, with neural activity in the right hemisphere no greater for the experimental task than the baseline. However, in most cases, there is at least some right-hemisphere activity, even if it is weaker than that in the left hemisphere.

78 Neuroimaging studies indicating right-hemisphere activation during various language tasks are reviewed in M. Jung-Beeman, "Bilateral Brain Processes for Comprehending Natural Language," *Trends in Cognitive Science* 9 (2005): 512–18. One key paper demonstrated increased right-temporal-lobe activity when people had to compute the theme or gist of a story: M. St. George et al., "Semantic Integration in Reading: Engagement of the Right Hemisphere During Discourse Processing," *Brain* 122 (1999): 1317–25.

Other papers demonstrate right-hemisphere involvement when people draw inferences, including R. A. Mason and M. Just, "How the Brain Processes Causal Inferences in Text: A Theoretical Account of Generation and Integration Component Processes Utilizing Both Cerebral Hemispheres," *Psychological Science* 15 (2004): 1–7; and S. Virtue, T. Parrish, and M. Jung-Beeman, "Inferences During Story Comprehension: Cortical Recruitment Affected by Predictability of Events and Working Memory Capacity," *Journal of Cognitive Neuroscience* 20 (2008): 2274–84.

Still other papers demonstrate right-hemisphere involvement in understanding jokes: S. Coulson and R. F. Williams, "Hemispheric Differences and Joke Comprehension," *Neuropsychologia* 43 (2005): 128–41; and V. Goel and R. J. Dolan, "The Functional Anatomy of Humor: Segregating Cognitive and Affective Components," *Nature Neuroscience* 4 (2001): 237–38.

78 A number of cases have now documented recovery from language loss due to left-hemisphere stroke accompanied by increased activity in right-hemisphere homologues to the damaged left-hemisphere language areas: S. S. Blank, "Speech Production After Stroke: The Role of the Right Pars Opercularis," *Annals of Neurology* 54 (2003): 310–20; V. Blasi et al., "Word Retrieval Learning Modulates Right Frontal Cortex in Patients with Frontal Damage," *Neuron* 36 (2002): 159–70; A. Leff et al., "A Physiological Change in the Homotopic Cortex Following Left Posterior Temporal Lobe Infarction," *Annals of Neurology* 51 (2002): 443–558; C. K. Thompson, "Neuroplasticity: Evidence from Aphasia," *Journal of Communication Disorders* 33 (2000): 357–66; C. K. Thompson and D. B. den Ouden, "Neuroimaging and Recovery of Language in Aphasia," *Current Neurological Neuroscience Reports* 8 (2008): 475–83; L. Winhuisen et al., "Role of the Contralateral Inferior Frontal Gyrus in Recovery of

Language Function in Poststroke Aphasia: A Combined Repetitive Transcranial Magnetic Stimulation and Positron Emission Tomography Study," *Stroke* 36 (2005): 1759–63.

It Takes All Kinds

79 Anatomical asymmetries, particularly at the neuronal level, are reviewed in J. Hutsler and R. A. W. Galuske, "Hemispheric Asymmetries in Cerebral Cortical Networks," *Trends in Neuroscience* 26 (2003): 429–36; and M. Jung-Beeman, "Bilateral Brain Processes for Comprehending Natural Language," *Trends in Cognitive Science* 9 (2005): 512–18. More detailed descriptions can be found in C. Chiarello et al., "A Large-Scale Investigation of Lateralization in Cortical Anatomy and Word Reading: Are There Sex Differences?" *Neuropsychology* 23 (2009): 210–22; R. Jacob, M. Schall, and A. B. Scheibel, "A Quantitative Dendritic Analysis of Wernicke's Area in Humans: II. Gender, Hemispheric, and Environmental Factors," *The Journal of Comparative Neurology* 327 (1993): 97–111; A. B. Scheibel et al., "Differentiating Characteristics of the Human Speech Cortex: A Quantitative Golgi Study," in *The Dual Brain: Hemispheric Specialization in Humans,* ed. D. F. Benson and E. Zaidel (New York: Guilford Press, 1985), 65–74; H. L. Seldon, "Structure of Human Auditory Cortex. II. Cytoarchitectonics and Dendritic Distributions," *Brain Research* 229 (1981): 277–94; J. Semmes, "Hemispheric Specialization: A Possible Clue to Mechanism," *Neuropsychologia* 6 (1968): 11–26; D. M. Tucker, D. L. Roth, and T. B. Blair, "Functional Connections Among Cortical Regions: Topography of EEG Coherence," *Electroencephalography Clinical Neurophysiology* 63 (1986): 242–50.

79 Steven Kosslyn and colleagues showed how coarse visual coding in the right hemisphere could explain aspects of visual information processing: Kosslyn et al., "Categorical Versus Coordinate Spatial Relations: Computational Analyses and Computer Simulations, *Journal of Experimental Psychology: Human Perception and Performance* 18 (1992): 562–77. Kosslyn and colleagues eventually rejected this idea based on subsequent findings: S. M. Kosslyn et al., "Hemispheric Differences in Sizes of Receptive Fields or Attentional Biases?," *Neuropsychology* 8 (1994): 139–47. However, we believe that this topic has not been thoroughly explored and that their rejection of coarse coding as an explanatory mechanism for right-hemisphere visual processing was premature. Moreover, it is possible that the distinction between coarse and fine coding occurs in association areas of the brain (areas without direct inputs or

outputs) but not in primary perceptual or motor areas. The hemispheric differences in neurons discussed in this chapter are well established in such association areas but not in primary perceptual areas.

79 The serendipitous findings of longer connections within the right rather than left temporal cortex are reported in E. Tardif and S. Clarke, "Intrinsic Connectivity of Human Auditory Areas: A Tracing Study with Dil," *European Journal of Neuroscience* 13 (2001): 1045–50.

Right On!

80 Mark's visual half-field studies of insight in the hemispheres with Edward Bowden are reported in M. Jung-Beeman and E. M. Bowden, "The Right Hemisphere Maintains Solution-Related Activation for Yet-to-Be Solved Insight Problems," *Memory and Cognition* 28 (2000): 1231–41; Bowden and Jung-Beeman, "Getting the Right Idea: Semantic Activation in the Right Hemisphere May Help Solve Insight Problems," *Psychological Science* 6 (1998): 435–40; Bowden and Jung-Beeman, "Aha! Insight Experience Correlates with Solution Activation in the Right Hemisphere," *Psychonomic Bulletin and Review* 10 (2003): 730–37.

CHAPTER 7: TUNING OUT AND GEARING UP

83 John Lasseter and Pete Docter's tour of Pixar Animation Studios can be found in the bonus material of disk 2 of the DVD *Monsters, Inc.*

The Idling Brain

84 In describing the role of alpha, we used the convenient analogy of idling. However, this analogy misses an important property of this type of brain activity. Alpha doesn't just idle a part of the brain. Rather, it inhibits or suppresses it so that the region won't interfere with activity elsewhere.

84 The discovery of the pre-insight "brain blink" was originally reported in M. Jung-Beeman et al., "Neural Activity Observed When People Solve Verbal Problems with Insight," *PLoS Biology* 2 (2004): 500–510, www.plosbiology.org/article/info%3Adoi%2F10.1371%2Fjournal.pbio.0020097/. A recent report has replicated our alpha brain blink using a different type of problem: L. Wu et al., "How Perceptual Processes Help to Generate New Meaning: An EEG Study of Chunk Decomposition in Chinese Characters," *Brain Research* 1296 (2009): 104–12.

Blinking the Mind's Eye

85 For background on Hans Berger, see *Wikipedia,* s.v. "Hans Berger," last modified July 7, 2014, http://en.wikipedia.org/wiki/Hans_Berger.

87 Aaron Sorkin has often talked about taking showers to stimulate his creative process: www.hollywoodreporter.com/news/aaron-sorkin-hbo-drama-news room-570176. Jonathan Franzen's techniques are described in an article in *The New York Times Magazine:* www.nytimes.com/2001/09/02/magazine/ jonathan-franzen-s-big-book.html.

Insight and Outsight

88 Our neuroimaging study of preparation for insight is described in J. Kounios et al., "The Prepared Mind: Neural Activity Prior to Problem Presentation Predicts Subsequent Solution by Sudden Insight," *Psychological Science* 17 (2006): 882–90. Additional discussion of preparation for insight and the role of the anterior cingulate can be found in J. Kounios and M. Beeman, "The Cognitive Neuroscience of Insight," *Annual Review of Psychology* 65 (2014): 71–93.

CHAPTER 8: THE INCUBATOR

92 Sir Paul McCartney has described on CNN's *Larry King Live* how the melody of the song "Yesterday" came to him in a dream: www.youtube.com/ watch?v=lLXq5Qc9zB4. McCartney further comments on songwriting in another segment from the same interview: www.youtube.com/watch ?v=062Mz-y3hk.

Hotbed of Incubation

93 The story of Otto Loewi's dream insight has been recounted in many sources. A summary can be found in *Wikipedia*, s.v. "Otto Loewi," last modified July 20, 2014, en.wikipedia.org/wiki/Otto_Loewi.

94 The story of Descartes's fly has been described many times, for example in B. R. Hergenhahn, *An Introduction to the History of Psychology* (Belmont, Calif.: Wadsworth, 2008), 118.

94 Matthew Walker provides a recent overview of the cognitive and affective neuroscience of sleep: M. P. Walker, "The Role of Sleep in Cognition and Emotion," *Annals of the New York Academy of Sciences,* vol. 1156, *The Year in Cognitive Neuroscience 2009,* 168–97.

Snooze, Lose, and Remember

95 The Pete Hamill quotes are drawn from an interview he gave on the National Public Radio program *Fresh Air* on May 5, 2011: www.npr.org/2011/05/05/135985200/pete-hamill-revisits-the-newsroom-in-tabloid-city.

Psychoanalyst Leanne Domash described another type of sleep-induced creative catharsis. Over a century ago, Freud noted that psychoanalysts sometimes dream about their patients. He believed that analysts absorb information about their patients at an unconscious level and that dreams are sometimes the vehicle of choice for the unconscious mind to convey helpful insights about the patient to the analyst. Though modern scientific psychology has not embraced Freud's view, Domash recounted an intriguing dream that she understood to reveal a therapeutically important insight about one of her patients. She had been treating a young woman for anxiety and depression. Over the course of several months of therapy, the patient had portrayed her mother as a "seductive, yet periodically cruel" woman. However, in spite of months of therapeutic discussions, Domash described herself as unable to "feel" it—Domash believed that she didn't truly understand the emotions about the patient's mother.

One night, Domash was awakened "shaken and sweating" by a dream containing a "chilling and dreadful" image. It was of a mermaid with the face of actress Joan Crawford. In her early years, the Academy Award winner had gained renown as a powerful actress with a number of screen classics to her credit. Nowadays, Crawford is perhaps most remembered as the focus of a notorious tell-all memoir by her daughter, *Mommie Dearest* (later made into an equally notorious motion picture), in which Crawford was depicted as a physically and emotionally abusive tyrant. After experiencing the horrifying image of a mermaid with Crawford's face, Domash felt that she finally understood her patient's emotions. This led her to realize that her relationship with her patient was in danger of drifting into a sadomasochistic reflection of the relationship between Crawford and her daughter. She attributed her realization to sleep's purported ability to remove blockages that prevent ideas from reaching awareness. Armed with this new perspective, Domash carefully steered the course of therapy away from this potential snare.

Domash's dream insight is described in Domash, "Unconscious Freedom and the Insight of the Analyst: Exploring Neuropsychological Processes Underlying 'Aha' Moments," *Journal of the American Academy of Psychoanalysis and Dynamic Psychiatry* 38 (2010): 315–40.

One of the few rigorous studies of incubation during sleep was recently done by Ut Na Sio and colleagues, who showed that sleep helps people solve the most difficult remote associates problems, but not easier ones: U. N. Sio, P. Monaghan, and T. Ormerod, "Sleep on It, but Only If It Is Difficult: Effects of Sleep on Problem Solving," *Memory and Cognition* 41 (2013). Unfortunately, their study didn't distinguish between insight and analytic solutions, so it doesn't show whether sleep provided a special boost to insight.

95 The study of the effects of sleep on the processing of memories is described in J. M. Ellenbogen et al., "Human Relational Memory Requires Time and Sleep," *Proceedings of the National Academy of Sciences of the USA* 104 (2007): 7723–28.

Some interesting studies directly suggest the existence of problem incubation during sleep, although they aren't definitive. Nevertheless, these studies are worth noting, because they show the trajectory that research on this topic is taking. For example, Gregory White examined a technique called "dream incubation," which some psychotherapists teach their patients to use in an effort to get them to tap the purported ability of the sleeping brain to solve problems. He asked his participants to pick one of their problems as their focus for the ten days of the experiment. White then randomly assigned these participants to one of four groups. One group was instructed to spend a few minutes thinking about a personal problem just before going to sleep each night. Other participants thought about their problem upon awakening in the morning. Two additional groups were instructed to spend a few minutes relaxing before sleeping or upon awakening, with no instructions to think about any problem.

After ten days, the participants who pondered their problem just before sleeping later rated the problem as more solvable and less distressing than the groups of participants who had relaxed or thought about their problem in the morning. This study is intriguing evidence that sleep made these participants feel better about their problems. Nevertheless, the study was ambiguous, primarily because it only asked participants how they felt about their problems rather than objectively determining whether they had actually made progress in solving them.

Another tantalizing study was described in an article entitled "Sleep Inspires Insight" by Ulrich Wagner and colleagues. Their participants were given a dreary cognitive task to perform, requiring them to substitute particular digits for other digits in a list of digit sequences using a rule supplied by the experimenter. What the participants weren't told was that there was another rule—a hidden shortcut—that would allow them to skip part of the way through each digit sequence and jump right to the last digit of that sequence, saving time and effort. Wagner defined insight as occurring whenever a participant became aware of the hidden rule and started skipping to the end of each digit sequence. Some were able to do this during the first session. Others weren't. But a night's sleep seemed to help participants to discover the hidden rule during the second session.

Wagner's study, though fascinating, must be considered tentative because of some problematic features. For example, the participants who discovered the hidden rule after sleep also slowed down their digit substitution performance before finding the hidden rule. So, there is no way to know whether sleep facilitated the discovery of the shortcut, or whether sleep just relaxed these participants enough for them to slow down and notice it. Most important, the researchers didn't determine whether awareness of the hidden rule came suddenly, which is the hallmark of insight, or whether it was gradual. This means that they may not have been studying the same process that both researchers and nonscientists typically think of as insight.

There are other uncertain aspects of the Wagner study. For example, participants weren't actually told to look for a hidden rule. By sticking with the rule initially given to them, they may have thought that they were doing what they were supposed to do. So in this case, their supposed insight wasn't so much the solution to a problem as it was the unmotivated realization that there was an alternate solution that they had the freedom to use. Some of the participants might even have achieved this realization right at the beginning of the experiment but thought that by using the hidden rule they were not following instructions. On the other hand, after sleeping, participants might simply have been less willing to perform the boring task and were actively looking for a shortcut that would reduce the tedium.

Sleep isn't homogenous. Researchers have isolated a number of distinct stages of sleep based on the patterns of neural activity present in each. A more general distinction is between two more-famous types of sleep: REM (rapid eye movement) and NREM (non-rapid eye movement). NREM sleep exhibits slow EEG waves and scant dream content; REM sleep exhibits faster EEG waves that look more like those occurring during waking consciousness and contains much more vivid dream content. For the typical night of sleep, the first half has more NREM than REM, while the second half has more REM than NREM.

Denise Cai, Sara Mednick, and colleagues recently investigated the potentially different roles that REM and NREM sleep may play in problem incubation. They gave their participants a series of remote associates problems to solve, then a series of analogy problems, and then a break. Some participants relaxed during the break, while others took a ninety-minute nap. After the break, the participants tried again to solve the remote associates problems that they had been unable to solve before the break. What they didn't know was that the analogy problems that they had seen earlier contained words that were solutions to the remote associates problems. Thus, the analogies surreptitiously primed participants to solve the remote associates problems after the break. The key finding was that participants whose breaks consisted of predominantly REM naps, rather than NREM naps, were better able to solve the remote associates problems that had been primed by the analogy words. The researchers interpreted this to mean that there is something special about REM sleep that helped these participants incubate the problems and make use of the primes.

Though these results suggest an intriguing role for REM sleep in incubation, they remain ambiguous. The design of the experiment does not allow a person to draw the conclusion that REM sleep caused superior performance on the primed problems, because the experimenters didn't *control* how much REM sleep each napper had. They just measured REM and sorted each participant into either the REM group or the NREM group. Instead of REM sleep causing an improvement in performance by fostering incubation, perhaps there was something else about these REM nappers—let's say, creativity—that made them better able to make use of the primes to solve the remote associates problems. Perhaps creative people are both more easily primed to incubate *and* are more likely to have REM sleep during a nap. According to this scenario, REM sleep could have been a by-product, and not the cause, of whatever it is that made these people effective creative problem solvers. So this study proposes an interesting idea for further research but doesn't yet allow us to conclude that REM is a neural manifestation or facilitator of incubation during sleep.

White's study of dream incubation is described in G. L. White and L. Taytroe, "Personal Problem-Solving Using Dream Incubation: Dreaming, Relaxation, or Waking Cognition?," *Dreaming* 13 (2003): 193–209. Wagner's study of the effects of sleep on insight: U. Wagner et al., "Sleep Inspires Insight," *Nature* 427 (2004): 352–55. Cai's study of REM sleep and the incubation of remote associates problems is described in D. J. Cai et al., "REM, Not Incubation, Improves Creativity by Priming Associative Networks," *Proceedings of the National Academy of Sciences of the USA* 106 (2009): 10130–34.

Sleep has yet other benefits. It not only consolidates your memories of the past, but it also strengthens your ability to remember to do things in the future, such as taking your medicine or running an errand. An unsolved problem is also a deferred task. The failure to solve it can sensitize you to hints around you that could trigger a solution by insight. Recent research suggests that sleep can heighten opportunistic assimilation of clues. The effects of sleep on prospective memory are described in M. K. Scullin and M. A. McDaniel, "Remembering to Execute a Goal: Sleep on It!," *Psychological Science* 21 (2010): 1028–35.

Waking Incubation

98 Sio and Ormerod's meta-analysis of past incubation research is described in U. N. Sio and T. C. Ormerod, "Does Incubation Enhance Problem Solving?: A Meta-Analytic Review," *Psychological Bulletin* 135 (2009): 94–120.

Fatigued Brains, Automatic Thinking, and the Real Unconscious Mind

98 Research on "attention restoration theory" backs up the idea that the brain's executive functions can be rejuvenated by even relatively short periods of rest. See, for example, M. Berman, J. Jonides, and S. Kaplan, "The Cognitive Benefits of Interacting with Nature," *Psychological Science* 19 (2008): 1207–12.

99 A flurry of recent studies on the unconscious thought hypothesis was sparked by an article by Dijksterhuis and colleagues: A. Dijksterhuis et al., "On Making the Right Choice: The Deliberation-Without-Attention Effect," *Science* 311 (2006): 1005–7. However, subsequent failures to reproduce such phenomena have cast the extreme version of the unconscious thought idea into doubt. For example, B. R. Newell and T. Rakow, "Revising Beliefs About the Merits of Unconscious Thought: Evidence in Favor of the Null Hypothesis," *Social Cognition* 29 (2011): 711–26.

Unfinished Business

102 Zeigarnik's work on memory for unfinished tasks is described in B. Zeigarnik, "Untersuchungen zur Handlungs- und Affektpsychologie: III. Das Behalten erledigter und unerledigter Handlungen" [Investigations on the psychology of action and affection: III. The memory of completed and uncompleted actions], *Psychologische Forschung* 9 (1927): 1–85; and B. Zeigarnik, "On Finished and Unfinished Tasks," in *A Sourcebook of Gestalt Psychology,* ed. W. D. Ellis (New York: Humanities Press, 1938), 300–314.

102 Seifert and colleagues describe opportunistic assimilation in M. C. Seifert et al., "Demystification of Cognitive Insight: Opportunistic Assimilation and the

Prepared-Mind Perspective," in *The Nature of Insight,* ed. R. J. Sternberg and J. E. Davidson (Cambridge, Mass.: MIT Press, 1995), 65–124.

103 Edward Bowden's aha moment at the opera is described in a National Public Radio story, www.npr.org/templates/story/story.php?storyId=1838162, and on page 568 of the following research report (which also describes his hint study): E. M. Bowden, "The Effect of Reportable and Unreportable Hints on Anagram Solution and the Aha! Experience," *Consciousness and Cognition* 6 (1997): 545–73.

104 Dorothy Wordsworth's account of the Wordsworths' walking tours and her account of "The Circumstances of 'Composed Upon Westminster Bridge'" are found in the *Grasmere Journals* from *Journals of Dorothy Wordsworth,* 2nd ed., ed. Mary Moorman (London: Oxford University Press, 1988). The Westminster account appears on page 436.

105 For information about Richard James and the Slinky, see *Wikipedia,* s.v. "Richard James," last modified February 4, 2014, en.wikipedia.org/wiki/Richard_T._James.

Get Over It!

105 Erik Verlinde's insight about gravity and the circumstances under which he had his aha moment are described in www.nytimes.com/2010/07/13/science/13gravity.html?adxnnl=1&adxnnlx=1285362013-syjvuZLrhSV4iTr0NG2xbA.

106 Vul and Pashler's study did not find incubation in the absence of misleading hints. This is probably because their participants worked on a number of problems before the incubation period, thereby "blurring" together incubation activity from several problems. Other research by Ben Baird, Jonathan Schooler, and colleagues mentioned in the last chapter of this book does show incubation without misleading hints when the incubation interval follows a single unsolved problem. Vul and Pashler's study is described in E. Vul and H. Pashler, "Incubation Is Helpful Only When People Are Misled," *Memory and Cognition* 35 (2007): 701–10. See also B. Baird et al., "Inspired by Distraction: Mind Wandering Facilitates Creative Incubation," *Psychological Science* 23 (2012): 1117–22.

107 Jarrod Moss's study of the timing of hints is described in J. Moss, K. Kotovsky, and J. Cagan, "The Effect of Incidental Hints When Problems Are Suspended Before, During, or After an Impasse," *Journal of Experimental Psychology: Learning, Memory, and Cognition* 37 (2011): 140–48.

107 Delaney's study of daydreaming and forgetting is described in P. F. Delaney et

al., "Remembering to Forget: The Amnesic Effect of Daydreaming," *Psychological Science* 21 (2010): 1036–142.

State of the Art

108 Another excellent study of waking incubation was described in a recent paper by Ben Baird and colleagues: B. Baird et al., "Inspired by Distraction: Mind Wandering Facilitates Creative Incubation," *Psychological Science* 23 (2012): 1117–22. They found superior incubation during an easy secondary task compared with either a demanding secondary task or no secondary task. So far, there is no definitive explanation for this fascinating finding. We touch on this in Chapter 14.

109 The Erroll Garner quotation is derived from www.playpiano.com/wordpress/pianists/erroll-garner-one-of-a-kind-jazz-pianist-composer-of-misty.

CHAPTER 9: IN THE MOOD

111 The authors interviewed Jerry Swartz on December 22, 2008.

113 Alexander Graham Bell's insight is described in many sources, including an interesting article by Malcolm Gladwell: www.newyorker.com/reporting/2008/05/12/080512fa_fact_gladwell.

113 Art Fry's idea for Post-it Notes is described in *Wikipedia,* s.v. "Arthur Fry," last modified July 21, 2014, en.wikipedia.org/wiki/Arthur_Fry. The Mozart quote appears in the Amabile article cited below.

From the Lab to the Workplace

114 For a review of Alice Isen's groundbreaking research on mood and creativity, including a theory of potential brain mechanisms, see F. G. Ashby, A. M. Isen, and U. Turken, "A Neuropsychological Theory of Positive Affect and Its Influence on Cognition," *Psychological Review* 106 (1999): 529–50. Amabile's research on workplace creativity is described in T. M. Amabile et al., "Affect and Creativity at Work," *Administrative Science Quarterly* 50 (2005): 367–403.

Crosstalk

116 Mood congruency is discussed in K. Fiedler, "Affective Influences on Social Information Processing," in *Handbook of Affect and Social Cognition,* ed. J. P. Forgas (Mahwah, N.J.: Lawrence Erlbaum Associates, 2001), 163–85.

Focus

118 A classic article describing the effects of mood on visual attention is J. A. Easterbrook, "The Effect of Emotion on Cue Utilization and the Organization of Behaviour," *Psychological Review* 66 (1959): 183–201. Gasper and Clore's study of mood and the spread of attention is described in K. Gasper and G. L. Clore, "Attending to the Big Picture: Mood and Global Versus Local Processing of Visual Information," *Psychological Science* 13 (2002): 34–40.

Illuminating Ideas

119 Anderson and colleagues' study of the effects of mood on perceptual and conceptual attention is described in G. Rowe, J. B. Hirsh, and A. K. Anderson, "Positive Affect Increases the Breadth of Attentional Selection," *Proceedings of the National Academy of Sciences USA* 104 (2007): 383–88. The study of the effects of mood on the N400 response is described in K. D. Federmeier et al., "Effects of Transient, Mild Mood States on Semantic Memory Organization and Use: An Event-Related Potential Investigation in Humans," *Neuroscience Letters* 305 (2001): 149–52. Barbara Fredrickson discusses the role of mood and attention in complex situations in B. Fredrickson and C. Branigan, "Positive Emotions Broaden the Scope of Attention and Thought-Action Repertoires," *Cognition and Emotion* 19 (2005): 313–32.

Marrying Head and Heart

122 Much of our research on mood and insight is described in K. Subramaniam et al., "A Brain Mechanism for Facilitation of Insight by Positive Affect," *Journal of Cognitive Neuroscience* 21 (2009): 415–32. Michael Posner discusses the role of the anterior cingulate in self-regulation in M. I. Posner et al., "The Anterior Cingulate Gyrus and the Mechanism of Self-Regulation," *Cognitive, Affective and Behavioral Neuroscience* 7 (2007): 391–95.

123 We should note that fMRI scans did not show greater activity in any part of the right hemisphere than the left during preparation for insight, providing no support for the idea that a positive mood can "turn on" the right hemisphere. But when a person is in a positive mood, the anterior cingulate is more sensitive to weakly activated remote associations, and such associations are mainly processed in the right hemisphere. So a positive mood may not turn on the right hemisphere, but it does set the stage for processes that will eventually occur.

Irreconcilable Differences

124 Many of the ideas in this section are based on K. Subramaniam et al., "A Brain Mechanism for Facilitation of Insight by Positive Affect," *Journal of Cognitive Neuroscience* 21 (2009): 415–32.

Repercussions of the Creative High

125 The idea that the relation between positive mood and associative processing is reciprocal is presented in M. Bar, "A Cognitive Neuroscience Hypothesis of Mood and Depression," *Trends in Cognitive Sciences* 13 (2009): 456–63. See also T. T. Brunyé et al., "Happiness by Association: Breadth of Free Association Influences Affective States," *Cognition* 127 (2013): 93–98; and M. F. Mason and M. Bar, "The Effect of Mental Progression on Mood," *Journal of Experimental Psychology: General* 141 (2012): 217.

Fluctuating Moods, Alternating Perspectives

126 Nancy Andreasen's work on creativity and psychopathology is summarized in N. C. Andreasen, *The Creating Brain: The Neuroscience of Genius* (New York: Dana Press, 2005). The idea that depression may adaptively foster rumination is presented in P. W. Andrews and J. A. Thomson, Jr., "The Bright Side of Being Blue: Depression as an Adaptation for Analyzing Complex Problems," *Psychological Review* 116 (2009): 620–54.

One idea is that the volume of ideas produced is increased while in a manic state. In fact, some researchers have proposed that creative breakthroughs are largely a product of chance and high productivity—if you have millions of monkeys banging on typewriter keyboards for millions of years, eventually one of them will accidently type out *Hamlet*. During mania, a dramatic increase in the sheer number of ideas generated could make it more likely that at least one of these ideas is worthwhile.

Indeed, a person in a manic state can be astonishingly fluent and prolific. For example, George Frideric Handel composed his great oratorio *Messiah* in twenty-four days during what was apparently a manic episode. (A few of the pieces in *Messiah* include recycled material, but how many people could even *copy,* never mind *compose,* the whole score of *Messiah* in just twenty-four days?)

Chance and productivity probably play a role, but they can't be the most important factors underlying creativity. Random production of ideas would be too slow. If, say, one idea in a thousand is really good, then a person would have to generate a thousand ideas, give or take, before stumbling on a good one. Every one of the fifty-three pieces in Handel's *Messiah* is masterful and contains outstanding musical ideas. Even in a manic state, could Handel have *randomly* generated, in less than three weeks, the many thousands of musical ideas necessary before he could accumulate enough to create fifty-three great compositions? For that matter, random idea generation absolutely cannot account for expert musical improvisation, because high-level improvisation cannot tolerate any bad ideas. Rather than productivity causing creativity, the reverse makes more sense—creativity makes productivity possible.

Finally, it's important to keep in mind that your moods don't completely dictate your cognitive style. Unless a mood is extreme, it gently nudges, rather than shoves, your brain to think about a situation in a particular way. It is possible to resist or circumvent this influence. For example, great athletes are admired for their resourcefulness in the clutch, as during the split-second decisions Brett Favre made while being chased by three-hundred-pound linebackers. And people like Wag Dodge, the Mann Gulch firefighter (Chapter 2), are capable of creative insight in even more extreme circumstances.

Science hasn't yet revealed how some people are able to do this. One possibility is that, in Dodge's case, with the fire racing toward him, he simply resigned himself to death. Speculatively, this state of acceptance could have freed him from terror-induced tunnel vision, thus enabling his lifesaving insight. More generally, perhaps those rare individuals who are capable of creativity under great stress are not exceptions to the principle that negative emotion impedes insight. It could be that their ability to distance themselves from a crisis and achieve a degree of inner calm enables them to be creative in spite of the threat.

CHAPTER 10: YOUR BRAIN KNOWS MORE THAN YOU DO

129 W. W. Morgan's quote is derived from Morgan, "A Morphological Life," *Annual Review of Astronomy and Astrophysics* 26 (1988): 1–9.

129 Rebecca Woodings's story can be found here: www.washingtonpost.com/wp-dyn/content/article/2009/12/14/AR2009121402863.html.

Thoughts from the Fringe

131 The quote from William Rowan Hamilton's letter to his son is adapted from en.wikiquote.org/wiki/William_Rowan_Hamilton. Information about

Hamilton can be found at *Wikipedia*, s.v. "William Rowan Hamilton," last modified July 28, 2014, en.wikipedia.org/wiki/William_Rowan_Hamilton. A relatively accessible explanation of Hamilton's idea is given at plus.maths.org/content/curious-quaternions. The picture of the plaque commemorating Hamilton's discovery is found at en.wikipedia.org/wiki/File:William_Rowan_Hamilton_Plaque_-_geograph.org.uk_-_347941.jpg.

The tip-of-the-tongue (TOT) phenomenon is another example of thought from the fringe. It sometimes occurs while you're trying to remember something, usually a word. As with intuition, you can't remember the specific word, but you know it's there, because it feels like it's on the tip of your tongue. Often, you might even know what letter the word starts with and how many syllables it has. If it's the name of a person, you may even be able to bring to mind an image of the person's face. If it's the name of an actor that you're trying to recall, you might even remember what movies she appeared in or some tabloid gossip about her. But her name remains elusive until it eventually pops into awareness. When this happens, like a sneeze, it relieves the discomfort.

Subjectively, the TOT phenomenon seems a lot like the kind of intuition that sometimes precedes insight, but there are differences. TOT states tend to become more common as a person ages, likely reflecting older people's increasing difficulty in remembering words. However, intuition doesn't seem to deteriorate with age. Another difference is that the TOT phenomenon is a difficulty in remembering a word that's otherwise well known to you. In contrast, intuition is the anticipation of a creative act—the emergent insight reflects a novel idea or perspective rather than a familiar one. At present, it isn't known whether intuition has any underlying similarity to the TOT phenomenon other than the fact that both seem to be tense anticipatory states preceding the sudden emergence of information into awareness. Research on the TOT phenomenon is reviewed in A. S. Brown, "A Review of the Tip-of-the-Tongue Experience," *Psychological Bulletin* 109 (1991): 204–23.

Intuition in the Laboratory

134 Arthur Reber's studies of unconscious processing are described in Reber, *Implicit Learning and Tacit Knowledge: An Essay on the Cognitive Unconsciousness* (New York: Oxford University Press, 1993).

Flying Blind

136 Janet Metcalfe's studies of "feelings of warmth" during problem solving are described in J. Metcalfe and D. Wiebe, "Intuition in Insight and Noninsight Problem Solving," *Memory and Cognition* 15 (1987): 238–46.

Inkling

138 Kenneth Bowers's studies of semantic coherence are described in K. Bower et al., "Intuition in the Context of Discovery," *Cognitive Psychology* 22 (1990): 72–110. Bolte and Goschke's studies of semantic coherence are described in A. Bolte, T. Goschke, and J. Kuhl, "Emotion and Intuition: Effects of Positive and Negative Mood on Implicit Judgments of Semantic Coherence," *Psychological Science* 14 (2003): 416–21; and A. Bolte and T. Goschke, "Intuition in the Context of Object Perception: Intuitive Judgments Rest on the Unconscious Activation of Semantic Representations," *Cognition* 108 (2008): 608–16.

Moody and Fluent; The Face of Intuition: Keep Smiling; *and* The Heavy Hand of Thought

140 Sascha Topolinksi and Fritz Strack's studies of fluency, mood, and intuition are described in S. Topolinski and F. Strack, "The Analysis of Intuition: Processing Fluency and Affect in Judgments of Semantic Coherence," *Cognition and Emotion* 23 (2009): 1465–503; and S. Topolinski and F. Strack, "The Architecture of Intuition: Fluency and Affect Determine Intuitive Judgments of Semantic and Visual Coherence and Judgments of Grammaticality in Artificial Grammar Learning," *Journal of Experimental Psychology: General* 138 (2009): 39–63. The story of Jackie Larsen is described on pages 31 and 32 of G. Myers, *Intuition: Its Power and Perils* (New Haven, Conn.: Yale University Press, 2002).

The Meaning of Life

148 Joshua Hicks's studies of mood and faith in intuition are described in J. A. Hicks et al., "Positive Affect, Intuition, and Feelings of Meaning," *Journal of Personality and Social Psychology* 98 (2010): 967–79. Amitai Shenhav's study of intuition and faith in God is described in A. Shenhav, D. G. Rand, and J. D. Greene, "Divine Intuition: Cognitive Style Influences Belief in God, *Journal of Experimental Psychology: General* 141 (2012): 423–28.

A Different Kind of Smarts

151 Scott Barry Kaufman's study of individual differences in intuitive ability is described in S. B. Kaufman et al., "Implicit Learning as an Ability," *Cognition* 116 (2010): 321–40.

Additional Notes on Intuition and Neuroimaging

In previous chapters, we discussed the role of a brain area called the "anterior cingulate." A number of studies have implicated the anterior cingulate in detecting weakly activated or competing pieces of information in other parts of the brain and guiding further processing based on the presence of this information. Our studies have shown that the anterior cingulate is prominently activated when a person prepares to solve a problem that is expected to be presented but hasn't yet appeared. But this happens only for problems that she will eventually solve with insight and not for problems she will solve analytically. We've argued that the anterior cingulate plays a key role in the preparation or readiness to detect weakly activated, unconscious ideas and guide further processing by either focusing on an unconscious idea or by suppressing it because it's a distraction. Evidence comes from R. Ilg et al., "Neural Processes Underlying Intuitive Coherence Judgments as Revealed by fMRI on a Semantic Judgment Task," *NeuroImage* 38 (2007): 228–48. Ilg and colleagues measured brain activity while their participants performed a semantic coherence task with word triads and found that when participants judged a triad to be coherent, the anterior cingulate was prominently active (compared with activation for incoherent triads). Functional magnetic resonance imaging studies by Kirsten Volz and colleagues sought to identify brain areas involved in intuition during visual and auditory perception: K. G. Volz, R. Rübsamen, and D. Y. von Cramon, "Cortical Regions Activated by the Subjective Sense of Perceptual Coherence of Environmental Sounds: A Proposal for a Neuroscience of Intuition," *Cognitive, Affective, and Behavioral Neuroscience* 8 (2008): 318–28; and K. G. Volz and D. Y. von Cramon, "What Neuroscience Can Tell About Intuitive Processes in the Context of Perceptual Discovery," *Journal of Cognitive Neuroscience* 18 (2006): 2077–87. They found that when participants made coherence judgments of visual and auditory stimuli (for example, whether a picture is a scrambled image of a real object, or whether a sound is a scrambled version of a familiar sound), another area of the brain called the rostral medial orbitofrontal cortex also became active. They proposed that this area, which lies just behind the eyes, forms an initial "quick hunch" and then uses the hunch to guide slower processing of the stimulus in the right temporal lobe. Interestingly, they note that this brain area has also been implicated in the processing of emotion, which is consistent with the findings of Topolinski, Strack, and others, indicating that intuition is closely related to changes in mood. At this point, however, it remains unclear whether the role of the rostral medial orbitofrontal cortex in intuition lies more in the formation of a hunch or in the processing of emotions associated with the formation of the hunch.

CHAPTER 11: THE INSIGHTFUL AND THE ANALYST

154 The quotes from Judah Folkman are derived from an interview for the PBS television series *NOVA*: www.pbs.org/wgbh/nova/cancer/folkman.html and from a commencement address given at Oberlin College in 2002: www.pbs.org/wgbh/nova/sciencenow/dispatches/080522.html. Additional material is adapted and quoted from Cooke, *Dr. Folkman's War,* 54–55, 88.

Different Strokes; Taking It All In; Doing the Right Thing; *and* Going Rogue

155 The EEG study described in this chapter is based on J. Kounios et al., "The Origins of Insight in Resting-State Brain Activity," *Neuropsychologia* 46 (2008): 281–91, and on unpublished data from the same study. The EEG coherence results are presented in J. Kounios and M. Jung-Beeman, "The Origins of Insight in Resting-State Brain Activity," a paper presented at the Meeting of the Psychonomic Society, Boston, November 2009.

A number of studies have reported that creative people have relatively unfocused attention. The following two recent articles summarize older research: P. I. Ansburg and K. Hill, "Creative and Analytic Thinkers Differ in Their Use of Attentional Resources," *Personality and Individual Differences* 34 (2003): 1141–52; and S. H. Carson, J. B. Peterson, and D. M. Higgins, "Decreased Latent Inhibition Is Associated with Increased Creative Achievement in High-Functioning Individuals," *Journal of Personality and Social Psychology* 85 (2003): 499–506.

157 The relationship between perceptual and conceptual attention is discussed by G. Rowe, J. B. Hirsh, and A. K. Anderson, "Positive Affect Increases the Breadth of Attentional Selection," *Proceedings of the National Academy of Sciences USA* 104 (2007): 383–88.

160 A number of studies have shown the relative heritability and stability over

time of a person's pattern of EEG, for example E. R. John et al., "Neurometrics: Computer-Assisted Differential Diagnosis of Brain Dysfunctions," *Science* 239 (1988): 162–69; D. J. A. Smit et al., "Heritability of Background EEG Across the Power Spectrum," *Psychophysiology* 42 (2005): 691–97; D. J. Smit et al., "Individual Differences in EEG Spectral Power Reflect Genetic Variance in Gray- and White-Matter Volumes," *Twin Research and Human Genetics* 15 (2012): 384.

A recent study in John's lab, not yet published, shows that the resting-state EEG predicts a tendency to solve problems insightfully, even when the EEGs were recorded several days before participants tackled the problems. This shows that the insight-related components of resting-state EEG tend toward stability.

Maddening Creativity

160 Historical information relevant to the story of Sir Isaac Newton and the apple can be found in Steve Connor, "The Core of Truth Behind Sir Isaac Newton's Apple," *The Independent,* January 18, 2010. The quote from Humphrey Newton, Isaac Newton's secretary and copyist, can be found on page 406 of S. Westphall, *Never at Rest: A Biography of Isaac Newton* (Cambridge, U.K.: Cambridge University Press, 1980). The quote from James Gleick came from an episode of the BBC Television series *Horizon* entitled "Isaac Newton: The Last Magician." The quote by Milo Keynes about Sir Isaac Newton's personality is from M. Keynes, "The Personality of Isaac Newton," *Notes and Records: The Royal Society Journal of the History of Science* 49 (1995): 1–56.

Despite his enormous contributions to science and mathematics, Newton's main interests lay elsewhere. When his unpublished manuscripts were finally made available to the general public in the twentieth century, it became clear that most of his writings were concerned with occult knowledge, especially alchemy and prophecy. The economist John Maynard Keynes purchased some of Newton's alchemical manuscripts in 1942. After reading them, he said that "Newton was not the first of the age of reason, he was the last of the magicians."

Just a Little Off

162 The current state of knowledge about schizophrenia is reviewed in the following four articles: R. Tandon, M. S. Keshavan, and H. A. Nasrallah, "Schizophrenia, 'Just the Facts': What We Know in 2008, Part 1: Overview," *Schizophrenia Research* 100 (2008): 4–19; Tandon, Keshavan, and Nasrallah,

"Schizophrenia, 'Just the Facts': What We Know in 2008, Part 2. Epidemiology and Etiology," *Schizophrenia Research* 102 (2008): 1–18; Tandon et al., "Schizophrenia, 'Just the Facts': What We Know in 2008, Part 3: Neurobiology," *Schizophrenia Research* 106 (2008): 89–107; and Tandon, Keshavan, and Nasrallah, "Schizophrenia, 'Just the Facts,' Part 4: Clinical Features and Conceptualization," *Schizophrenia Research* 110 (2009): 1–23.

163 A general overview of schizotypy is presented by M. F. Lenzenweger in "Schizotypy: An Organizing Framework for Schizophrenia Research," *Current Directions in Psychological Science* 15 (2006): 162–66.

164 Just as there are different types of schizophrenia, there are apparently also different types of schizotypy. Only "positive" schizotypy, which is characterized by unusual thoughts and experiences, is thought to be related to enhanced creativity and right-hemisphere functioning: J. E. Fisher, et al., "Neuropsychological Evidence for Dimensional Schizotypy: Implications for Creativity and Psychopathology," *Journal of Research in Personality* 38 (2004): 24–31.

164 The study showing that schizotypes are better at solving insight problems than are non-schizotypes was reported by A. Karimi et al., "Insight Problem Solving in Individuals with High Versus Low Schizotypy," *Journal of Personality Research* 41 (2007): 473–80.

164 The overinclusive thought of schizotypes is discussed by C. Mohr et al., "Loose but Normal: A Semantic Association Study," *Journal of Psycholinguistic Research* 30 (2001): 475–83.

Getting It On and Getting Along

165 The evolutionary persistence of schizophrenia genes and evidence that schizotypes engage in more sexual activity than non-schizotypes is discussed by D. Nettle and H. Clegg, "Schizotypy, Creativity and Mating Success in Humans," *Proceedings of the Royal Society B* 273 (2006): 611–15.

166 The quotations from Ed Catmull, president of Pixar, are derived from an interview at a conference sponsored by *The Economist* in March 2010 in New York City: www.scottberkun.com/blog/2010/inside-pixars-leadership/.

Uninhibited Thought

168 Research on negative priming and schizotypy is reviewed by R. K. Minas and S. Park, "Attentional Window in Schizophrenia and Schizotypal Personality: Insight from Negative Priming Studies," *Applied and Preventive Psychology* 12 (2007): 140–48.

168 For a study suggesting that alcohol can increase insight solutions for remote associates problems, see A. F. Jarosz, G. J. Colflesh, and J. Wiley, "Uncorking the Muse: Alcohol Intoxication Facilitates Creative Problem Solving," *Consciousness and Cognition* 21 (2012): 487–93.

Your Finest Hour

169 For time-of-day effects in problem solving, see M. B. Wieth and R. T. Zacks, "Time of Day Effects on Problem Solving: When the Non-Optimal Is Optimal," *Thinking and Reasoning* 17 (2011): 387–401. See also C. P. May, "Synchrony Effects in Cognition: The Costs and a Benefit," *Psychonomic Bulletin and Review* 6 (1999): 142–47.

Self-Control

169 Ezra Wegbreit's study of the persistence of attentional focus and its effects on insight is described in E. Wegbreit et al., "Visual Attention Modulates Insight Versus Analytic Solving of Verbal Problems," *The Journal of Problem Solving* 4 (2012): 6.

Weighty Insights

171 Some additional thoughts from the cutting-room floor: Our discussion suggests that the brains of schizotypes are different from non-schizotypes in important ways. In Chapter 6, we discussed how the two cerebral hemispheres process information differently. In a typical right-handed person, a concept strongly activates a small number of closely associated ideas in the left hemisphere and a larger number of more remotely associated ones in the right hemisphere. The right hemisphere thus makes a unique contribution to the insight process through its processing of remote associations.

Besides processing close associations, the left hemisphere is usually also dominant for basic types of language processing. However, in schizotypes and schizophrenics, this left-hemisphere language dominance is reduced, or even reversed. This likely explains schizotypes' loose associative thought and greater incidence of left-handedness and ambidexterity. (The right hemisphere controls the left hand.) It seems that schizotypes' right hemispheres play a larger role than those of non-schizotypes.

How far can this idea be pushed? Our research has compared Insightfuls' and Analysts' brain activity and found that Insightfuls have more active right hemispheres. Is this greater reliance on the right hemisphere fleeting or relatively stable? Ongoing research in John's lab suggests that it is relatively stable. Future studies of brain anatomy can also help to answer this question. Brain structure can change, but slowly. If a more active right hemisphere is structurally different from a less active one, then this would suggest that right hemisphere prominence is fairly stable.

A comparison of the brain structures of Insightfuls and Analysts has yet to be done. However, recent research suggests that there are structural differences in hemispheric dominance that could bias people to be habitually insightful or analytical. Bruce Morton and Stein Rafto used a variety of tests to determine which hemisphere was dominant in 149 test subjects. They examined MRI scans of these subjects' brains, focusing on a part of the anterior cingulate. Remember that the anterior cingulate plays an important role in monitoring and controlling processes throughout the brain and that our research has shown the involvement of this area in preparation for insight. Their astonishing finding was that in 98 percent of their test subjects, a particular part of the anterior cingulate averaged 50 percent thicker in each person's dominant hemisphere than in the nondominant hemisphere. They theorized that the brain has a single "executive observer" that monitors and guides brain activity. It resides in the side of the anterior cingulate that is thicker. Thus, the executive observer's hemisphere dominates. Morton and Rafto's work thus supports the idea that differences in cognitive style originate in relatively stable differences in brain structure.

There are, however, a couple of caveats that should be made. Here we have a perfect example of the old chicken-and-egg problem. Which comes first? Does a person tend to be a "left-brained thinker" or a "right-brained thinker" because the anterior cingulate or some other structure is larger or thicker on that side? Or is it the case that thinking in a left-brained or a right-brained way enlarges key brain areas in that hemisphere in the same way that using a muscle makes it both stronger and larger?

Even if future research does support the idea that differences in cognitive style are due to physical asymmetries in the brain, this doesn't mean that even those individuals with the most asymmetric brains are forever locked into one style of thought. Just because one hemisphere may be dominant doesn't mean that the other one is totally submissive. As we've seen, the hemispheres contribute to thought by working together seamlessly. One's genes may influence the size and thickness of brain structures, but intensive use or training—what cognitive psychologists call "deliberate practice"—can modify the brain. Perhaps how people tend to think causes such differences in brain structure. If so, then training could change aspects of hemispheric dominance. In sum, even if cognitive style is determined by one's brain anatomy and genes, that doesn't imply that one's experience and training have no effect.

Morton and Rafto's findings of hemispheric asymmetry are discussed in B. E. Morton and S. E. Rafto, "Behavioral Laterality Advance: Neuroanatomical Evidence for the Existence of Hemisity," *Personality and Individual Differences* 49 (2010): 34–42.

CHAPTER 12: CARROTS AND STICKS

The Paradox of Motivation

175 Glucksberg's classic study of the effects of financial incentives on performance in solving the Candle Problem is described in S. Glucksberg, "The Influence of Strength of Drive on Functional Fixedness and Perceptual Recognition," *Journal of Experimental Psychology* 63 (1962): 36–41. Figure 12.1 is taken from upload.wikimedia.org/wikipedia/commons/d/d6/Genimage.jpg.

Zooming In and Locking On

177 Some of Harmon-Jones's studies of attention and motivation are described in E. Harmon-Jones and P. A. Gable, "Neural Activity Underlying the Effect of Approach-Motivated Positive Affect on Narrowed Attention," *Psychological Science* 20 (2009): 406–9; E. Harmon-Jones and C. K. Peterson, "Supine Body Position Reduces Neural Response to Anger Evocation," *Psychological Science* 20 (2009): 1209–10; and E. Harmon-Jones and P. Gable, "The Blues Broaden, but the Nasty Narrows: Attentional Consequences of Negative Affects Low and High in Motivational Intensity," *Psychological Science* 21 (2010): 211–15.

The Approaching Paradox

179 Some of Förster and Friedman's research on the effects of approach and avoidance motivation on cognitive styles is reported in J. Förster et al., "Enactment of Approach and Avoidance Behavior Influences the Scope of Perceptual and Conceptual Attention," *Journal of Experimental Social Psychology* 42 (2006): 133–46.

Types of Motivation

180 E. Tory Higgins's theory of regulatory focus is summarized on his website: www.columbiá.edu/cu/psychology/higgins/research.html.

Prizes and Pink Slips

180 The website for the XPRIZE Foundation can be found at www.xprize.org/. Information about Jack Welch's executive practices can be found at *Wikipedia,* s.v. "Jack Welch," last modified July 18, 2014, en.wikipedia.org/wiki/Jack_Welch.

Ready or Not

182 Teresa Amabile's research on deadlines and creativity is summarized in an accessible fashion in T. Amabile, C. N. Hadley, and S. J. Kramer, "Creativity Under the Gun," *Harvard Business Review* 80 (2002): 52–61. Our work on deadlines and insight is described in R. W. Smith and J. Kounios, "Sudden Insight: All-or-None Processing Revealed by Speed-Accuracy Decomposition," *Journal of Experimental Psychology: Learning, Memory, and Cognition* 22 (1996): 1443–62; and J. Kounios et al., "The Origins of Insight in Resting-State Brain Activity," *Neuropsychologia* 46 (1996): 281–91.

CHAPTER 13: FAR, DIFFERENT, UNREAL, CREATIVE

Symbols

186 Symbolic priming effects on insight are discussed in J. Förster et al., "Automatic Effects of Deviancy Cues on Creative Cognition," *European Journal of Social Psychology* 35 (2005): 345–59; and M. L. Slepian et al., "Shedding Light on Insight: Priming Bright Ideas," *Journal of Experimental Social Psychology* 46 (2010): 696–700.

Love and the Time Traveler

188 Förster's studies of mental time travel are described in J. Förster, R. S. Friedman, and N. Liberman, "Temporal Construal Effects on Abstract and Concrete Thinking: Consequences for Insight and Creative Cognition," *Journal of Personality and Social Psychology* 87 (2004): 177–89. Förster's study of the ef-

fects of love and lust on creativity is described in J. Förster, K. Epstude, and A. Ozelsel, "Why Love Has Wings and Sex Has Not: How Reminders of Love and Sex Influence Creative and Analytic Thinking," *Personality and Social Psychology Bulletin* 35, 1479–91.

What If . . . ?

189 The effects on creativity of counterfactual thinking are described in K. D. Markman et al., "Implications of Counterfactual Structure for Creative Generation and Analytical Problem Solving," *Personality and Social Psychology Bulletin* 33 (2007): 312–24.

Creativity from Afar

190 Multicultural effects on insight are described in W. W. Maddux and A. D. Galinsky, "Cultural Borders and Mental Barriers: The Relationship Between Living Abroad and Creativity," *Journal of Personality and Social Psychology* 96 (2009): 1047–61.

Distant Thoughts

192 Trope and Liberman's theory of psychological distance is described in Y. Trope and N. Liberman, "Construal-Level Theory of Psychological Distance," *Psychological Review* 117 (2010): 440–63. Jia's study showing that insight problems thought to be composed in a distant research institute were solved more accurately than when the same problems were thought to be composed in a nearby institute is described in L. Jia, E. R. Hirt, and S. C. Karpen, "Lessons from a Faraway Land: The Effect of Spatial Distance on Creative Cognition," *Journal of Experimental Social Psychology* 45 (2009): 1127–31.

The Big Picture

194 To ponder "what spring is like on Jupiter or Mars" with the Chairman of the Board: www.youtube.com/watch?v=K_ib4DRYflM&feature=fvst.

CHAPTER 14: THE STATE

197 A recent study by Jennifer Mueller shows people's resistance to creative leaders: J. S. Mueller, J. A. Goncalo, and D. Kamdar, "Recognizing Creative Lead-

ership: Can Creative Idea Expression Negatively Relate to Perception of Leadership Potential?," *Journal of Experimental Social Psychology* 47 (2011): 494–98.

197 The saying "People will accept your ideas much more readily if you tell them Benjamin Franklin said it first" is floating around the Internet and is usually attributed to someone named David H. Comins. There seems to be little or no information available about Comins except for the hypothesis that he is a real estate broker: c-pol.blogspot.com/2010/07/who-in-world-is-david-h-comins .html.

Training the Mind

199 The story about the slow elevators seems to have been told and adapted many times: www3.sympatico.ca/karasik/GF_evolution_of_legend.html. We thank Dr. Joseph Armenti for calling this story to our attention.

200 Tony McCaffrey's insight training procedure is described in T. McCaffrey, "Innovation Relies on the Obscure: A Key to Overcoming the Classic Problem of Functional Fixedness," *Psychological Science* 23 (2012): 215–18.

200 For a study of insight training that shows generalization to other types of problems, see J. E. Davidson and R. J. Sternberg, "The Role of Insight in Intellectual Giftedness," *Gifted Child Quarterly* 28 (1984): 58–64. However, one recent study suggests that insight training may benefit the solving of puzzle-like problems more than realistic problems: J. B. Cunningham and J. N. MacGregor, "Training Insightful Problem Solving: Effects of Realistic and Puzzle-Like Contexts," *Creativity Research Journal* 20 (2008): 291–96.

Prodding the Brain

203 Chi and Snyder's studies of insight and transcranial direct current stimulation are described in R. P. Chi and A. W. Snyder, "Facilitate Insight by Non-Invasive Brain Stimulation," *PLoS ONE* 6 (2011); R. P. Chi and A. W. Snyder, "Brain Stimulation Enables the Solution of an Inherently Difficult Problem," *Neuroscience Letters* 515 (2012), 121–24.

Obviously, no one should ever apply an electric current to any part of the human body without the approval and direct supervision of both a scientist and medical doctor trained in this technique and with the institutional oversight and safety controls of a hospital or research university.

Making It Happen

204 Most of the points summarized in this section are discussed in detail in earlier chapters. The appropriate citations can be found in the notes for those chapters. Below are citations for studies mentioned in this chapter but not in earlier chapters.

A recent study shows that high ceilings induce people to think more abstractly: J. Meyers-Levy and R. Zhu, "The Influence of Ceiling Height: The Effect of Priming on the Type of Processing That People Use," *Journal of Consumer Research* 34 (2007): 174–86.

Recent studies show that the colors blue and green, which are reminiscent of the expansive sky, ocean, and green fields, expand thought and enhance creativity; red, which is associated with blood, warning lights, and fire engines, constricts thought and enhances analysis: S. Lichtenfeld, A. J. Elliot, M. A. Markus, and P. Reinhard, "Fertile Green Facilitates Creative Performance," *Personality and Social Psychology Bulletin* 38 (2012): 784–797. R. Mehta and R. Zhu, "Blue or Red? Exploring the Effect of Color on Cognitive Task Performances," *Science* 323 (2009): 1226–29.

Studies show that sharp edges and other visual features grab attention by implying threat: M. Bar and M. Neta, "Humans Prefer Curved Visual Objects," *Psychological Science* 17 (2006): 645–48; and C. L. Larson et al., "Recognizing Threat: A Simple Geometric Shape Activates Neural Circuitry for Threat Detection," *Journal of Cognitive Neuroscience* 21 (2009): 1523–35.

People think more abstractly in dark surroundings: A. Steidle, L. Werth, and E.-V. Hanke, "You Can't See Much in the Dark: Darkness Affects Construal Level and Psychological Distance," *Social Psychology* 42 (2011): 174–84.

A case for self-experimentation along with interesting examples can be found in S. Roberts, "Self-Experimentation as a Source of New Ideas: Ten Examples About Sleep, Mood, Health, and Weight," *Behavioral and Brain Sciences* 27 (2004): 227–62.

Ben Baird, Jonathan Schooler, and colleagues recently showed that incubation yields superior results during an easy secondary task: B. Baird et al., "Inspired by Distraction: Mind Wandering Facilitates Creative Incubation," *Psychological Science* 23 (2012): 1117–22.

Enemies of the State

215 Raichle's ideas about the default-state mode are summarized in an accessible fashion in M. Raichle, "The Brain's Dark Energy," *Scientific American,* March

2010, 44–49. Two recent fMRI studies investigated the relationship between the default-state network and mental travel. Spreng and Grady showed how different forms of mental simulation tap the same default-state network of brain areas: R. N. Spreng and C. L. Grady, "Patterns of Brain Activity Supporting Autobiographical Memory, Prospection, and the Theory of Mind, and Their Relationship to the Default Mode Network," *Journal of Cognitive Neuroscience* 22 (2009): 1112–23. Tamir and Mitchell extended these results, also showing stable individual differences in the predilection to think about psychologically distant, rather than near, things: D. I. Tamir and J. P. Mitchell, "The Default Network Distinguishes Construals of Proximal Versus Distal Events," *Journal of Cognitive Neuroscience* 23 (2011): 2945–55.

The Path Forward

218 The story of Helen Keller's insight triggered by Anne Sullivan's touch is derived from Keller's autobiography. The quotes about Keller by Winston Churchill and Mark Twain are derived from Wikiquote: en.wikiquote.org/wiki/Helen_Keller.

INDEX

Note: Page numbers in *italics* refer to illustrations.

ABOUT THE AUTHORS

JOHN KOUNIOS, PH.D., is a professor of psychology and director of the doctoral program in Applied Cognitive and Brain Sciences at Drexel University. He has held research and faculty positions at Princeton University, the University of Pennsylvania, and the National Center for Post-Traumatic Stress Disorder. He is a fellow of the Association for Psychological Science and the Psychonomic Society. His research has been funded by grants from the National Science Foundation and the National Institutes of Health. He lives with his wife and children in the Philadelphia area.

MARK BEEMAN, PH.D., is chair and professor of psychology at Northwestern University, studying the brain bases of creative cognition and problem solving, how mood affects attention and cognition, and how the right and left sides of the brain differ in function. His research has been funded by the National Institutes of Health, the National Science Foundation, the John Templeton Foundation, and the Office of Naval Research. He is a Kavli fellow of the National Academy of Sciences and a fellow of the Association for Psychological Science. He is constantly learning from his teenage daughter and son.